Analysis of Variance via Confidence Intervals

Kevin D. Bird

SAGE Publications
London • Thousand Oaks • New Delhi

First published 2004

SAGE Publications Ltd
1 Oliver's Yard
55 City Road
London EC1Y 1SP

SAGE Publications Inc
2455 Teller Road
Thousand Oaks, California 91320

SAGE Publications India Pvt Ltd
Post Box 4109
B-42 Panchsheel Enclave
New Delhi 110 017

British Library Cataloguing in Publication data

A catalogue record for this book is available from the British Library.

ISBN 0 7619 6357 X

Library of Congress Control Number available

Printed in India by Gopsons Papers Ltd, Noida

Analysis of Variance via Confidence Intervals

Contents

Preface

In recent years a great deal of emphasis has been placed on the value of interval estimates of effect sizes in psychological and biomedical research. The International Committee of Medical Journal Editors (1997), the APA Task Force on Statistical Inference (Wilkinson and Task Force on Statistical Inference, 1999) and the fifth edition of the APA Publication Manual (2001) all recommend that confidence interval analysis should be preferred to null hypothesis significance testing.

Standard treatments of analysis of variance (ANOVA), the most widely used method of data analysis in experimental psychology, provide very little guidance about how a confidence interval approach can be implemented in the ANOVA context. When confidence intervals are discussed (usually very briefly), the purpose is often to illustrate their value in interpreting outcomes that are not statistically significant, rather than as the primary method of producing statistical inferences.

My purpose in writing this book has been to provide a treatment of fixed-effects ANOVA informed by Jason Hsu's (1996) hierarchy of levels of inference on comparisons. When applied to analyses based on ANOVA models, this hierarchy makes it clear that interval estimates of contrasts on effect parameters are more informative than (and imply) the lower-level inferences (directional, inequality and homogeneity inferences) produced by traditional approaches to ANOVA. The relationships between inferences at different levels in the hierarchy provide a basis for comparison between the analyses recommended here and the analyses typically reported in the literature.

The standard null hypothesis significance testing approach provides no satisfactory basis for the interpretation of nonsignificant outcomes. One of the most important advantages of the confidence interval approach is the basis it provides for the interpretation of contrasts and other functions of effect parameters (such as Cohen's f) when those parameters cannot be declared non-zero by a statistical test. If the experimenter is able to specify the smallest non-trivial value of an effect size parameter, an appropriate confidence interval will show whether the data can justify the claim that the parameter is trivially small, whatever the outcome of the corresponding significance test. I have included a number of examples of practical equivalence inference based on confidence intervals. The possibility of practical equivalence inference is particularly important in analyses of factorial designs, where the interpretation (or even the interpretability) of main effects is often deemed to depend on the assumption that interaction parameters are trivially small.

An approach to ANOVA emphasizing confidence interval inference on contrasts presents a particular challenge to students and researchers who rely on statistical packages to implement their analyses. Most of the popular statistical packages provide only limited support (if any) for confidence interval analyses, particularly contrast-based analyses compatible with traditional ANOVA F tests. I have dealt with this problem by providing access to *PSY*, a program that can produce most of the contrasts analyses discussed in this book. Some of these analyses (of data from single-factor between-subjects or within-subjects designs, or two-factor designs with one factor of each type) are particularly easy to implement with *PSY*. Others (particularly those requiring critical values of the greatest characteristic root or the studentized maximum root distributions) are currently difficult (if not impossible) to implement with most statistical packages.

I hope that this book will be useful as a text or reference for senior undergraduate and graduate students in psychology and related disciplines, and as a handbook for established researchers who want to learn how to implement and interpret confidence interval analyses in the ANOVA context. Although this is a small book relative to many others that deal with ANOVA models and procedures in any detail, it does deal with some 'advanced' topics (that is, topics typically not discussed in ANOVA textbooks). Among these are noncentral confidence intervals on standardized effect size parameters such as Cohen's d and f, nonstandard definitions of families of factorial contrasts to allow for the inclusion of simple effect contrasts in coherent analyses of factorial designs, and the MANOVA-model approach to the construction of simultaneous confidence intervals when an experimental design includes at least one within-subjects factor. I have shown how all of these analyses can be implemented with *PSY* or with popular statistical packages such as *SPSS*.

A number of people have contributed directly and indirectly to the production of this book and to the *PSY* program on which it substantially depends. Various versions of *PSY* have been used for teaching and research for some time in the School of Psychology at the University of New South Wales, and I would like to acknowledge the support for the further development of the program provided by Kevin McConkey, Sally Andrews and Peter Lovibond during their respective periods as Head of School. It would be difficult to overstate Dusan Hadzi-Pavlovic's contribution to the development of *PSY*. Dusan provided the programming that produces critical values of Boik's studentized maximum root distribution and Roy's greatest characteristic root distribution, thereby allowing *PSY* to produce simultaneous confidence intervals appropriate for a range of analyses based on product contrasts.

My understanding of various issues associated with ANOVA-model analyses has been sharpened by numerous discussions with Melanie Gleitzman, with whom I have shared the teaching of ANOVA to senior undergraduate psychology students at UNSW for more years than either of us cares to contemplate. I have also benefited from conversations and collaborations over a number of years with Dusan Hadzi-Pavlovic and Wayne Hall.

I am indebted to Liz Hellier and her colleagues for allowing me to use their data as a basis for most of the within-subjects analyses discussed in Chapter 6. This particular data set provides a good illustration of how practical equivalence inference is sometimes possible with repeated measures data, even without large sample sizes.

Finally, my thanks to Geoff Cumming for the thoughtful and detailed feedback he provided after reading earlier versions of a number of chapters. It was particularly useful to have the benefit of Geoff's experience in bringing the benefits of confidence interval analysis to the attention of students and researchers.

1 Comparing Two Means

Introduction

Analysis of variance (ANOVA) is the standard method used to generate confident statistical inferences about systematic differences between means of normally distributed outcome measures in randomized experiments. In order to provide a context for what follows, we will begin by considering briefly what is implied by the previous sentence.

First, the reference to *randomized* experiments suggests that ANOVA is particularly appropriate for the analysis of data from experiments in which each subject (participant, experimental unit) is randomly assigned to one of two or more different treatment conditions (experimental groups). All subjects in a particular treatment condition receive the same treatment, and differences in the effects of the treatments are expected to produce differences between groups in post-treatment scores on a relevant outcome measure. Random assignment of subjects to experimental conditions usually ensures that systematic differences between means of groups given different treatments can be attributed to differences in the effects of the treatments, rather than to any other explanation such as the presence of pre-existing differences between the groups. If nothing goes wrong in the conduct of a randomized experiment, systematic differences between group means on a dependent variable can quite properly be attributed to differences in the effects of the treatments, and ANOVA procedures are specifically designed to produce inferences about differential treatment effects.

ANOVA procedures are not well suited for the analysis of data from quasi-experiments, where different treatments are given to pre-existing groups, such as different classes in a school. It is usually preferable to analyse quasi-experimental data by methods that attempt to take pre-existing differences into account (Reichardt, 1979).

The statement that ANOVA (or at least fixed-effects ANOVA, by far the most widely used version) is concerned with the pattern of differences between *means* implies that, despite the name of the procedure, the focus is on means rather than variances.[1] When the outcome of an experiment is not well summarized by a set of means, as is the case when the dependent variable is categorical rather than continuous, ANOVA is not appropriate.

The theory underlying inferential procedures in ANOVA assumes (among other things) that the distributions to be summarized in terms of means are *normal* distributions. ANOVA procedures are often used, sometimes with very little justification, when outcome measures are not even approximately normally distributed.

We should also consider what is meant by *systematic* differences between means on an outcome measure. According to the simplest ANOVA model, the difference between two sample means (the means calculated from the data produced by subjects given two different treatments in a randomized experiment) is influenced by two independent components, one of which is systematic, the other random. The systematic component is the difference between the effects of the two treatments; in a simple randomized experiment this difference between effects can be thought of as a difference between two *population* means, one for each treatment. The experimenter's problem, of course, is that the observed difference between sample means can also be influenced by a number of other unknown factors, some of which are associated with the particular subjects who happen to be assigned to a particular treatment, and some of which may be regarded as factors contributing to measurement error. In a randomized experiment these unknown factors contribute to the difference between sample means in a random or nonsystematic way. ANOVA methods allow for the influence of random as well as systematic influences on sample means.

Finally, it is necessary to consider what is meant by *confident* statistical inference. Suppose that the dependent variable in a two-group randomized experiment is a 40 item ability test, and the sample means on this test are $M_1 = 25.76$ and $M_2 = 19.03$, so that the difference between sample means is $M_1 - M_2 = 6.73$. Given this difference, which is specific to this particular sample (or pair of samples), what can we infer about the difference between the effects of the two treatments (that is, the difference between population means $\mu_1 - \mu_2$)? Because $M_1 - M_2$ is an unbiased estimator of $\mu_1 - \mu_2$ it might seem reasonable to infer that $\mu_1 - \mu_2 = 6.73$. This is indeed the best *point* estimate of the value of $\mu_1 - \mu_2$. The problem with a point estimate of a difference between two population means is that it is almost certain to be different from the actual value of the parameter being estimated, so that we are almost certain to be wrong if we assert that $\mu_1 - \mu_2 = 6.73$. Imagine *replicating* the experiment (repeating the experiment with different subjects) a very large number of times.[2] Values of $M_1 - M_2$ would vary across replications, almost never being exactly equal (given an unlimited number of decimal places) to the unknown value of $\mu_1 - \mu_2$. Because we cannot be confident in any sense that this (or any other) point estimation procedure is likely to produce a correct inference, point estimation is not a confident inference procedure. An *interval estimate* of the value of $\mu_1 - \mu_2$, however, does allow for the possibility of confident inference.

Suppose that the 95% confidence interval (CI) on $\mu_1 - \mu_2$ turns out to have a lower limit of 4.39 and an upper limit of 9.07, and as a consequence we assert that the value of $\mu_1 - \mu_2$ is somewhere between these limits. We cannot be sure that this particular inference is correct, but we know that the CI procedure controls the probability (at .95) that inferences of this kind are correct, meaning that 95% of the CIs constructed in this way (one interval per replication) would include the value of $\mu_1 - \mu_2$. We can express the same thing in terms of an *error rate* by saying that the probability of being in error when asserting that the interval includes $\mu_1 - \mu_2$ is .05. (This means that 5% of assertions of this kind, one per replication, are incorrect.) The difference between confident statistical inference and point estimation is that the former allows the researcher to control the probability of inferential error.

ANOVA provides for confident inference on differences between means via CIs and statistical significance tests. As we will see, inference from CIs is more informative than inference from tests.

Organization of this book

This book is intended for readers who have completed at least one course in statistics and data analysis including a reasonably substantial treatment of statistical inference. If you feel the need to revise some of the basic ideas underlying statistical inference before dealing with the material in this book, you will find that the accounts given by Lockhart (1998) and Smithson (2000) are compatible with the approach taken here.

Chapter 1 applies a hierarchy of levels of inference to the problem of producing and justifying a confident inference on a single comparison between two means. If your introduction to statistical inference emphasized statistical hypothesis testing you may be surprised to discover that CI inference occupies the highest level in this hierarchy.

Chapter 2 deals with the application of the ANOVA model to data from single-factor between-subject designs. CI inference on individual contrasts (generalized comparisons) is emphasized, and the hierarchy of inference levels is used to show how this approach is related to traditional approaches emphasizing test-based heterogeneity and directional inference.

In Chapter 3 we consider methods of controlling the precision of CI inferences on contrasts. The relationship between precision of interval estimates on contrasts and the analogous concept of the power of significance tests on contrasts is discussed.

Chapter 4 deals with simple between-subjects factorial designs in which the effects of varying the levels of at least two different experimental factors are examined within a single experiment. In this chapter we examine a number of different approaches to the problem of producing coherent inferences on contrasts on three types of factorial effects defined by ANOVA models: simple effects, main effects and interaction effects. Analyses of complex between-subjects factorial designs are considered in Chapter 5.

In Chapter 6 we consider within-subjects (repeated measures) designs, where each subject in a single group is subjected to all of the treatments examined in an experiment, or where each subject is examined on a number of trials or measurement occasions. ANOVA models for within-subjects designs must somehow deal with the fact that repeated measurements on a single individual are not independent of one another, so these models, and the data-analytic procedures that make use of them, differ in important ways from the models and methods discussed in earlier chapters.

Chapter 7 deals with mixed factorial designs: designs with at least one between-subjects factor and at least one within-subjects factor.

Many of the analyses recommended in this book are not currently supported by most of the popular statistical packages. You can carry out most of these analyses with a program called *PSY* (Bird, Hadzi-Pavlovic and Isaac, 2000). See Appendix A for a general overview of the program and the website from which it can be downloaded. *PSY* does not carry out some of the more traditional analyses based on significance tests that are supported by statistical packages, and it does not carry out analyses based on various extensions of ANOVA models. For these reasons (among others), Appendix B provides *SPSS* syntax required to carry out various analyses with *SPSS*, probably the most popular of all statistical packages. Finally, some of the more advanced analyses are most easily carried out with the *STATISTICA Power Analysis* program (Steiger, 1999). The use of this program is discussed where appropriate.

The data sets used for most of the examples and exercises can be downloaded from the Sage website at

http://www.sagepub.co.uk/resources/bird.htm

Every input file mentioned in this book can be downloaded from that website and can be opened by *PSY*.

Confident inference on a single comparison

Many of the basic ideas concerning inference in ANOVA can be developed in the context of a two-group randomized experiment, where the experimenter wishes to use the data to produce a confident inference on a single comparison between two means. In this chapter we will examine in some detail the logic of confident inference in the two-group case. Suppose that an experimenter wishes to examine the effect of a treatment (an experimental manipulation of some kind) by comparing the mean score of treated subjects on a relevant dependent variable (outcome measure) with the mean score obtained by a different group of subjects who did not receive the treatment (a control group). N subjects are randomly assigned to two groups (with $n = N/2$ subjects per group), and steps are taken to ensure that the problems that can sometimes arise in randomized experiments (Cook and Campbell, 1979, Shadish, Cook and Campbell, 2002) do not arise in this one.

The parameter of most interest to the experimenter is $\mu_1 - \mu_2$, the difference between the mean of all potential subjects who receive the treatment and the mean of all potential subjects in the control condition. This difference between the means of two hypothetical populations is unknown, but can be estimated from $M_1 - M_2$, the difference between the means of subjects in the two groups. In a randomized experiment the *expected value* of $M_1 - M_2$ is $\mu_1 - \mu_2$. This means that if the experiment were replicated an infinite number of times and each replication produced a value of $M_1 - M_2$, then the mean of the distribution of $M_1 - M_2$ values would be $\mu_1 - \mu_2$. $M_1 - M_2$ is therefore an *unbiased* estimator of $\mu_1 - \mu_2$.

Absence of bias is a desirable property of an estimator, but it does not imply that the estimate of an effect size obtained from a single experiment will be so close to the actual effect size that the discrepancy between the two can be safely ignored. A confident inference about the value of $\mu_1 - \mu_2$ can be obtained from an analysis of variance, but in the two-group case the same inference can be obtained from the familiar two-group t test or from a t-based CI. It will be convenient to deal with a number of issues in this familiar context before we discuss the more general ANOVA model. We need to remember that the standard two-group t (test or CI) procedure is based on the assumption that both populations of dependent variable values have a normal distribution, and that the two within-population standard deviations are identical.[3]

Strength of inference on a comparison

Following Hsu (1996), we will distinguish between three levels of confident inference on a comparison between two means: CI inference, confident direction inference and confident inequality inference. These three levels are ordered in terms of strength of inference, CI inference being stronger than the other two.

Confidence interval inference In Hsu's terminology, which we will adopt here, a CI inference on the comparison $\mu_1 - \mu_2$ can be expressed as

$$\mu_1 - \mu_2 \in (ll, ul)$$

where the symbol '$\in$' means 'is included in' or 'is covered by',

ll is the lower limit of the interval,

and ul is the upper limit of the interval.

Thus the inference $\mu_1 - \mu_2 \in (10.1, 12.3)$ asserts that

$$10.1 < (\mu_1 - \mu_2) < 12.3,$$

that is, that μ_1 is greater than μ_2 by at least 10.1 units, but by no more than 12.3 units. The CI (as distinct from the inference derived from it) is (10.1, 12.3).

If the CI covers the parameter (that is, if $\mu_1 - \mu_2$ is in fact somewhere between the upper and lower limits), then the inference is correct, and no error has been made. If the interval does not cover the parameter (that is, if $\mu_1 - \mu_2$ is lower than the interval's lower limit or higher than the upper limit), then the inference is false. We will use the term *noncoverage error* to refer to this type of inferential error. Given the assumptions required to justify the CI procedure, the *noncoverage error rate* α (the probability of a noncoverage error) can be controlled at a nominated low level, usually set at .05 or .10. It is customary to specify the noncoverage error rate indirectly by setting the *confidence level*, defined as $100(1 - \alpha)\%$, at a high level. If the confidence level is set at 95%, then the noncoverage error rate is $\alpha = .05$.

It is important to understand what the confidence level and the noncoverage error rate mean, and also what they do not mean. Imagine that a two-group experiment is replicated a very large number of times. Each replication produces a 95% CI on $\mu_1 - \mu_2$. Because sample means and other statistics vary across replications, CI limits and the inferences following from them will also vary across replications. Provided that the relevant assumptions are satisfied, 95% of these repetitions of the experiment will produce a CI covering the population mean difference, thereby producing a correct inference. The (unknown) value of the parameter of interest $(\mu_1 - \mu_2)$ does not vary across these replications. The parameter is fixed, but the CIs vary across replications.

The language sometimes used to describe CIs ('the probability that the parameter lies inside the interval is ...') can encourage incorrect interpretations implying that the parameter is a variable rather than a constant. The important point is that the probability statement refers to the relative frequency with which variable intervals include (or exclude) the fixed parameter, given an indefinitely large number of replications of the experiment.

Confident direction inference A directional inference specifies the direction of the difference between means, but nothing more than that. Directional inference on the comparison $\mu_1 - \mu_2$ is an assertion of the form $\mu_1 > \mu_2$ or $\mu_1 < \mu_2$.

CI inference implies directional inference, provided that the CI excludes the value zero so that all values in the interval have the same sign. Thus the CI inference $\mu_1 - \mu_2 \in (10.1, 12.3)$ implies that $\mu_1 > \mu_2$.

In practice, a directional inference usually follows from the rejection of a null hypothesis such as $H_0: \mu_1 - \mu_2 = 0$ by a statistical test. Many test procedures designed to test hypotheses about differences between means (including the two-group t test) are associated with (and derivable from) a CI procedure. If a 95% CI constructed with the t procedure (the standard procedure in the two-group case) turns out to be (10.1, 12.3), justifying the directional inference $\mu_1 > \mu_2$, then the associated .05-level two-tailed t test will necessarily reject the null hypothesis $\mu_1 - \mu_2 = 0$. Rejection of this hypothesis implies the inequality inference $\mu_1 \neq \mu_2$, but does not by itself imply anything about the direction of the difference between means. The justification for directional inference in this case depends on the relationship between the CI and the test: the test is able to reject the null hypothesis when $M_1 > M_2$ if and only if the associated CI justifies the directional inference $\mu_1 > \mu_2$. Similarly, a statistically significant outcome when $M_1 < M_2$ implies that the associated CI includes only negative values [such as (–8.6, –2.3)], thereby justifying the directional inference $\mu_1 < \mu_2$.

We may note in passing that a two-sided CI (with both an upper and lower limit) implies the outcome of a two-tailed test, but it does not imply the outcome of a one-tailed test. One-tailed tests are associated with single-sided CIs (see Hsu, 1996), which are rarely used in psychological research. One disadvantage of single-sided CIs is that they provide no information about precision of estimation.

Erroneous directional inferences can occur under two conditions: first, if a directional inference is made when no difference exists; second, if a difference exists in the direction opposite to that asserted. If $\mu_1 - \mu_2 = 0$ and it is asserted that $\mu_1 > \mu_2$ or $\mu_1 < \mu_2$, then a Type I error has been made. If $\mu_1 - \mu_2 = 0$, then the Type I error rate from an α-level two-tailed test procedure is equal to the noncoverage error rate for confidence interval inference. The Type I error rate is hypothetical, in the sense that it refers to errors that cannot occur unless $\mu_1 = \mu_2$. If $\mu_1 > \mu_2$ and it is asserted that $\mu_1 < \mu_2$, or if $\mu_1 > \mu_2$ and it is asserted that $\mu_1 < \mu_2$, then a Type III error has been made. The Type III error rate cannot exceed $\alpha/2$, which it approaches for very small values of $|\mu_1 - \mu_2|$.

Confident inequality inference Inequality inference occurs when a data analyst asserts simply that $\mu_1 \neq \mu_2$, without specifying the direction of the difference between means. This is a particularly weak form of inference, because it denies only that the two means are absolutely identical. CI inference implies confident inequality inference if zero is excluded from the CI: the inference $\mu_1 - \mu_2 \in$ (10.1, 12.3) implies that $\mu_1 \neq \mu_2$. Confident direction inference also implies confident inequality inference, because the assertions $\mu_1 > \mu_2$ and $\mu_1 < \mu_2$ both imply that $\mu_1 \neq \mu_2$. A Type I error is the only type of error possible when inequality is asserted.

Equality inference? Following Hsu (1996), we do not consider Type II errors when defining error rates for directional inference or inequality inference. A Type II error (as defined in most discussions of significance tests) would occur if it were incorrectly asserted that $\mu_1 = \mu_2$. The approach we are considering makes no provision for such an inference. Indeed, a number of statisticians and data analysts argue that there is always some difference, perhaps an extremely small difference, between any two population means (Cohen, 1994, Schmidt, 1996). It is important, however, to allow for the possibility of inferring that two population means might be *practically equivalent*, meaning that the difference between them is in some sense trivially small. We will discuss practical equivalence inference after we consider effect size.

Interpreting effect size

CI inference is stronger (more informative) than confident direction or confident inequality inference, and is therefore emphasized in this book. Traditionally, however, the benefits of CI inference have been largely ignored in practice, and researchers have generally relied on significance tests that provide relatively weak inference, and no inference at all about the magnitude of the effects under investigation. Since the early 1960s, this traditional approach to inference has been severely criticized (see Harlow, Mulaik and Steiger, 1997, for a comprehensive review). No doubt the conservative approach taken by many textbook authors, journal editors and software producers has played a role in reinforcing the much criticized traditional approach. Researchers often respond to an obvious need to say something about the magnitude of an experimental effect by relying on a significance test to justify an assertion about the existence of the effect (via confident direction or confident inequality inference), then discussing the magnitude of the observed difference between means as though it is not subject to sampling error. As a consequence, a confident inference may (or may not) be made about the existence or direction of an effect, but no attempt is made to produce a confident inference about the magnitude of the effect. A much better approach is to incorporate the intention to make statements about effect size into the inferential analysis from the outset.

Before the magnitude of an experimental effect can be profitably estimated, it must be defined in a way that makes sense to the researcher. Although various approaches to effect size measurement have been proposed, a difference between two relevant population means has distinct advantages over alternative approaches, provided that the difference is expressed in an informative metric.

Dependent variable units The most obvious metric for an effect size is the metric used to scale the dependent variable. In well-developed areas of research, experimenters often have access to a substantial literature documenting effects of all sorts of experimental manipulations on dependent variables scaled in a common metric. Experiments in cognitive psychology, for example, often measure reaction time scaled in milliseconds. Researchers in this area should have little difficulty in deciding whether a difference between population means of a given magnitude (in units of milliseconds) should be regarded as trivially small, extremely large, or somewhere in between.

If a dependent variable is not scaled in informative units, then a CI scaled in dependent variable units will provide no useful inferences beyond confident direction inference. Suppose, for example, that a two-group randomized experiment is carried out to examine differences in the effects of two levels of blood alcohol on performance in a driving task. Subjects assigned to one group consume sufficient alcohol to raise their blood alcohol concentration (BAC) to 0.08%, while subjects in the second group have their BAC raised to 0.05%. Subjects are required to negotiate a slalom course designed specifically for the experiment, and the dependent variable is the number of cones (witches hats) hit by the car, a relatively low score indicating relatively good performance. It is expected, of course, that drivers with a BAC of 0.08 will hit more cones than will drivers with a BAC of 0.05. The experimenter is primarily interested in the size of that difference. The problem is that differences between means in the number of cones hit will be influenced by arbitrary features of this particular slalom, such as the difference between the width of the lane defined by the cones and the width of the car, the distance between adjacent cones, and the length of the slalom. The units of the dependent variable are essentially arbitrary, so an effect size expressed in these units may provide little or no useful information beyond the direction of the difference. Suppose that, unknown to (and unknowable by) the experimenter, the difference between population means is $\mu_{0.08} - \mu_{0.05} = 3.6$ cones. By itself, this figure conveys almost no useful information beyond the directional inequality $\mu_{0.08} > \mu_{0.05}$. The parameter value implies that a BAC of 0.08 increases the average number of cones hit, relative to a BAC of 0.05, but this may be a small and trivial effect, or a substantial and important effect. This is not an issue about inference from statistics to parameters; it is an issue about the interpretation of parameters.

Standard deviation units The conventional way of dealing with an effect size scaled in arbitrary units of measurement is to divide it by a relevant standard deviation scaled in the same arbitrary units, thereby removing the influence of those units. The resulting quantity is scaled in standard deviation units, thereby making it informative to anyone familiar with these units. Following Cohen (1965), it has become standard practice to assume that both population standard deviations are equal $(\sigma_1 = \sigma_2 = \sigma)$, so the standardized difference between two means (usually called Cohen's *d*) is $(\mu_1 - \mu_2)/\sigma$.[4] Cohen (1969) suggested that in the absence of any better basis for interpreting effect size, standardized effect sizes of 0.2, 0.5 and 0.8 should be interpreted respectively as small, medium and large effects. These suggestions have been widely accepted, and now have the status of established conventions.

Returning now to the slalom example, assume for the moment that the standard deviation of slalom scores in both populations (drivers with a BAC of 0.05 and drivers with a BAC of 0.08) is $\sigma = 20.3$ cones. (Incidentally, this means that the slalom must have a very large number of cones.) The mean difference of 3.6 cones is equivalent to a difference of 3.6 cones/20.3 cones = 0.18 standard deviation units. Most people who are familiar with standard deviation units would interpret this as a small and possibly trivial difference.

If the common population standard deviation were 3.3 cones (rather than 20.3 cones), then the standardized mean difference would be would be 1.09 standard deviation units, a large difference in terms of Cohen's guidelines.

While standardized effect sizes can sometimes provide a basis for relatively informative CI inference when none might otherwise exist, it should not be assumed that standardization always provides more information than the original scaling. A difference between means scaled in an interpretable dependent variable metric can be more informative than the corresponding standardized effect size, because the standardizing transformation removes information in this case. Returning again to the driving experiment, consider another test that might be used to assess the difference between the effects of the two blood alcohol levels. This test requires subjects to perform an emergency braking task: after an appropriate signal is given, the car is to be brought from a particular speed to a complete stop in the shortest possible distance. The dependent variable is braking distance: the distance travelled by the car after the presentation of the signal. It is expected that the higher blood alcohol level will produce longer braking distances than the lower level, but the experimenter is primarily interested in estimating the magnitude of the difference, to be scaled in metres, a familiar and informative unit of distance. The practical implications of an effect size expressed in units of metres can be discussed without any reference to standard deviation units.

Practical equivalence inference

If it is not possible to justify a confident inference that two treatments have identical effects on an outcome measure (that is, the inference $\mu_1 = \mu_2$), what can we make of claims that a treatment has no effect, or that a new treatment is no better than an old treatment, or that the effect of a treatment on males is the same as the effect on females? The problem with assertions of equality is not that there is something peculiar about asserting that the value of a parameter is zero, but rather that it is not possible to justify a confident inference about any point estimate. This problem can be resolved by replacing point estimation with interval estimation.

The first step in practical equivalence inference is to define a range of values of $\mu_1 - \mu_2$ within which the two treatments would be interpreted as equivalent for practical (or theoretical) purposes. In standard deviation units, as assessed by Cohen's (1969) effect size guidelines, the required difference might be very small, small or small–medium. The important point is that the experimenter, a group of researchers, a regulatory body or someone else must be able to specify the maximum difference that can be regarded as trivially small. Call this maximum trivially small difference τ. Then the two treatments are deemed to be practically equivalent if $\mu_1 - \mu_2 \in (-\tau, +\tau)$.

This *practical equivalence interval* defines what is to be meant by a trivially small effect. It is highly desirable, of course, that the value of τ should be specified independently of the data used to justify a claim of practical equivalence, and that researchers in a particular domain should be able to agree on what is to be meant by practical equivalence in that domain.

Note that the practical equivalence interval is a definition, not a CI or any other kind of statistical inference. It is possible to know what is meant by a trivially small effect without being confident that the actual difference between two population means is, in fact, trivially small.

If it turns out when an experiment is run that the CI on $\mu_1 - \mu_2$ lies entirely within the practical equivalence interval, then the inference from the CI implies that the two treatments must have practically equivalent effects on the dependent variable. If the agreed-upon practical equivalence interval is $(-0.25\sigma, 0.25\sigma)$ and it turns out that the CI from an experiment is $(-0.05\sigma, 0.17\sigma)$, then all of the values inside the CI are trivially small. Therefore the CI $\mu_1 - \mu_2 \in (-0.05\sigma, 0.17\sigma)$ implies that $\mu_1 \approx \mu_2$, where the symbol '$\approx$' means 'is practically equivalent to'.

Practical equivalence inference, then, is a special case of CI inference. It is not possible to justify a confident assertion about practical equivalence from a single inference at any lower level (such as confident direction inference or confident inequality inference).[5] It is possible to be confident about practical equivalence and also be confident about direction. The CI $(0.01\sigma, 0.18\sigma)$, for example, justifies the inference $\mu_1 > \mu_2$, and it also justifies the inference $\mu_1 \approx \mu_2$ if the practical equivalence interval is defined as $(-0.20\sigma, 0.20\sigma)$.

Practical equivalence inference is possible only from experiments providing a high degree of precision in estimation, as evidenced by narrow CIs relative to the practical equivalence interval. In practice, it turns out that most randomized experiments in psychology and related disciplines are not capable of producing sufficiently precise estimates of effect size to justify practical equivalence inference, whatever the outcomes of those experiments may be. Within-subjects (repeated measures) designs usually produce more precise estimates of effects than fully randomized between-subjects designs. Practical equivalence inference is sometimes possible from within-subjects designs, even when the sample size is not large. Chapter 6 contains examples of practical equivalence inference from a within-subjects design.

Practical equivalence inference is taken very seriously in some areas of research where the consequences of errors in claims of equivalence can be important, as might be the case when a relatively inexpensive new drug treatment is being considered as a replacement for an expensive standard treatment, and it is important to discover whether the two treatments have practically equivalent effects. In other areas, confident practical equivalence inference is sometimes possible on the basis of a meta-analysis (a quantitative analysis of results from a set of similar studies with a very large total sample size). The requirements of confident practical equivalence inference are the same in the context of meta-analysis as in any other context: a relevant CI must be included in (covered by) a relevant practical equivalence interval.

It is often difficult to justify a precise value of τ (the maximum trivially small difference), so the limits of a practical equivalence interval are often somewhat fuzzy. If the fuzziness is extreme, then practical equivalence inference is not possible.

Constructing a confidence interval on a single comparison

Population standard deviation known

Some of the basic principles about CI construction can be illustrated most clearly if we assume not only that dependent variable scores are normally distributed with the same standard deviation in each population, but also that the experimenter knows the value of the population standard deviation.

Given these assumptions, the procedure for constructing a raw $100(1 - \alpha)\%$ CI on $\mu_1 - \mu_2$ is quite straightforward. First, an unbiased point estimate of the difference between population means (namely $M_1 - M_2$) is calculated, together with the standard error (SE) of that estimate. Second, the standard error is multiplied by a critical value (CV) from a relevant theoretical probability distribution to determine the half-width (w) of the CI.[6] Finally, the CI limits are obtained by adding (and subtracting) the half-width to (and from) the estimated parameter value. That is, the limits of the interval are obtained from

$$(M_1 - M_2) \pm \text{CV} \times \text{SE}.$$

If both groups have the same sample size ($n_1 = n_2 = n = N/2$), the standard error of the difference between means is

$$\text{SE} = \sigma_{M_1-M_2} = \sigma\sqrt{\frac{2}{n}} \tag{1.1}$$

and the relevant critical value is CV $= z_{\alpha/2}$, the $100(1-\alpha/2)$th percentile point of the z (standard normal) distribution. The $100(1 - \alpha)$% CI is

$$\mu_1 - \mu_2 \in (M_1 - M_2) \pm z_{\alpha/2} \times \sigma_{M_1-M_2}. \tag{1.2}$$

An alternative and popular way of writing the same CI is

$$(M_1 - M_2) - z_{\alpha/2} \times \sigma_{M_1-M_2} < \mu_1 - \mu_2 < (M_1 - M_2) + z_{\alpha/2} \times \sigma_{M_1-M_2}.$$

In this book we will use expressions like (1.2).

Consider an experiment with n = 20 subjects in each of two groups, where the experimenter somehow knows before running the experiment that the population standard deviation is σ = 15. It follows from (1.1) that the standard error of the difference between the two sample means, a parameter whose value can also be known before the experiment is run, will be $15\sqrt{2/20} = 4.743$. The critical value required for a 95% CI is the 97.5th percentile point of the standard normal distribution, namely $z_{.025} = 1.960$. Therefore the half-width of a 95% CI constructed with the z procedure will be $w = 1.960 \times 4.743 = 9.30$.

Suppose that the sample means from the experiment turn out to be $M_1 = 105.31$ and $M_2 = 98.54$, so that the unbiased point estimate of $\mu_1 - \mu_2$ is $M_1 - M_2 = 6.77$. The limits of the 95% CI are obtained from $(M_1 - M_2) \pm w$ $= 6.77 \pm 9.30$. The required raw CI, then, is $\mu_1 - \mu_2 \in (-2.53, 16.07)$.

If a standardized CI is required and σ is known to be 15, then any of the statistics of interest can be transformed into standard deviation (SD) units by dividing by 15. The difference between sample means is $(6.77/15)\sigma = 0.45\sigma$ (0.45 SD units) and the CI can be expressed as $\mu_1 - \mu_2 \in (-0.17\sigma,\ 1.07\sigma)$, or in SD units as $(\mu_1 - \mu_2)/\sigma \in (-0.17,\ 1.07)$. The best point estimate of the standardized difference between population means suggests a medium effect. This estimate is not particularly precise, and all that can be inferred at the 95% confidence level is that either μ_1 is greater than μ_2 by some unknown but nontrivial amount (somewhere between small and large, but not massively large), or the two population means are practically equivalent. The interval does, therefore, justify the inference that μ_2 is not nontrivially larger than μ_1.

Precision The precision of inferences from this CI procedure can be determined before the experiment is run. The standard error of the difference between means is $\sigma\sqrt{2/n} = 0.316\sigma$, and the half-width of the CI is $1.960 \times 0.316\sigma = 0.620\sigma$. The experimenter knows in advance that if the difference between sample means turns out be zero, so that the data contain no suggestion of a difference between population means, the 95% CI will nevertheless include non-negligible positive values (such as $\mu_1 - \mu_2 = 0.6\sigma$), as well as non-negligible negative values (such as $\mu_1 - \mu_2 = -0.6\sigma$). Prior information about precision of estimation can be very useful: an experimenter who expects a small effect may decide to increase the sample size to increase precision, or, if the required resources are not available, to change the experimental design or even abandon the experiment altogether. If a very large effect is expected (such as $\mu_1 - \mu_2 = 2\sigma$), a CI half-width of 0.620σ might indicate an acceptable level of precision.

Implications for directional inference The fact that zero is inside the obtained CI implies that a two-tailed .05 level z test would not have rejected the null hypothesis H_0: $\mu_1 - \mu_2 = 0$, so no inference would have been justified by the test. (The z test rather than the t test is relevant here because of the assumption that σ is known.) In general, it is difficult to know what to make of the failure of a test to produce a directional or inequality inference. Either the difference between $\mu_1 - \mu_2$ and zero is trivially small, or the statistical power of the test has been insufficient to enable the test to detect whatever nontrivial difference exists. It is more difficult to determine the power of the test than it is to determine the precision of the CI, because the power of the test depends on the (unknown) magnitude of $\mu_1 - \mu_2$. It turns out that if $\mu_1 - \mu_2 = 0.8\sigma$ (a large effect according to Cohen's effect size guidelines), then the power of the test is 0.72; if $\mu_1 - \mu_2 = 0.5\sigma$ (a medium effect), then the power is 0.35. Figures like these can provide some help in the task of interpreting a 'nonsignificant' difference, but it is much easier to make sense of the associated CI.

In practice, the z method of CI construction is rarely used, because experimenters are rarely (if ever) in a position to know the population standard deviation. A value of $\sigma = 15$ might perhaps be assumed if the dependent variable is an IQ test scaled so that the standard deviation in the standardization population was set at 15. Even this case is problematic, however, unless subjects in the experiment are randomly sampled from the same population used to standardize the test. In general, it is not necessary to assume that the population standard deviation is known, because methods making using of t distributions do not require this assumption.

Population standard deviation unknown

We now abandon the assumption that the population standard deviation is known, but retain the assumption of normally distributed dependent variable scores with the same standard deviation in each population. Because σ is unknown, it is not possible to calculate the standard error of the difference between means. It is necessary to estimate σ from the data in order to use an expression similar to (1.1) to estimate $\sigma_{M_1 - M_2}$. ANOVA procedures produce a statistic based on variation within groups called *mean square error* (MS_E), an unbiased estimate of the population variance σ^2. We will defer a detailed discussion of this statistic until we discuss the ANOVA model in Chapter 2. At this point we merely note that it is possible to obtain an appropriate estimate $\hat{\sigma}^2$ of the unknown parameter σ^2, and we can use $\hat{\sigma} = \sqrt{\hat{\sigma}^2}$ to estimate σ. Replacing σ with $\hat{\sigma}$ in (1.1) produces an expression for the estimated standard error of the difference between sample means (a statistic), rather than the standard error itself (a parameter). If each group has n subjects, the estimated standard error is

$$\hat{\sigma}_{M_1-M_2} = \hat{\sigma}\sqrt{\frac{2}{n}}. \tag{1.3a}$$

If the two groups have different sample sizes (n_1 and n_2), the standard error is estimated from

$$\hat{\sigma}_{M_1-M_2} = \hat{\sigma}\sqrt{\frac{1}{n_1}+\frac{1}{n_2}}. \tag{1.3b}$$

Because we must estimate σ from the data in order to use (1.3a) or (1.3b), the critical value required to construct a CI is a percentile point from a t distribution rather than the standard normal distribution. The half-width of a raw $100(1-\alpha)\%$ CI is $t_{\alpha/2;N-2} \times \hat{\sigma}_{M_1-M_2}$, where $t_{\alpha/2;N-2}$ is the value of the upper $100(1-\alpha/2)$th percentile of the *central* t distribution with $(N-2)$ degrees of freedom. (A central t distribution is an 'ordinary' t distribution. We give it its full title here because we will subsequently need to distinguish between central and noncentral t distributions.) Therefore the CI is

$$\mu_1 - \mu_2 \in (M_1 - M_2) \pm t_{\alpha/2;N-2} \times \hat{\sigma}_{M_1-M_2}. \tag{1.4}$$

The t procedure produces exact $100(1-\alpha)\%$ raw CIs, in the sense that if the relevant assumptions are satisfied, the noncoverage error rate produced by the procedure is exactly α.

Suppose that the experiment under discussion produces a variance estimate of $\hat{\sigma}^2 = 234.78$ (slightly larger than the actual population variance of $\sigma^2 = 225$). An experimenter who does not know the value of σ^2 would use the estimated value to produce an estimated standard error of $\hat{\sigma}_{M_1-M_2} = \sqrt{234.78 \times 2/20} =$ 4.845. The experimenter would be in no position to know that in this particular case the estimated standard error is slightly larger than the actual value of $\sigma_{M_1-M_2} = 4.743$. The critical value required to construct a 95% CI is $t_{.025;38} =$ 2.024. (Note that this is slightly larger than the critical z value of 1.960 used previously when the experimenter supposedly knew the population standard deviation.) The half-width of the interval is $2.024 \times 4.845 = 9.81$, so the confidence limits are 6.77 ± 9.81, and the interval is (–3.04, 16.58).

This interval is wider than that calculated on the assumption that the experimenter knows the value of the population standard deviation. Two factors contribute to the difference in width. The first is the fact that when the standard deviation is unknown, CI width depends on a statistic (the estimated standard error), and therefore varies across samples. In this particular case the estimated standard error happens to be larger than the actual standard error, but the reverse could just as easily have been the case. The second factor contributing to the difference in CI width is the use of a critical value from a t distribution rather than the z distribution. A critical t value is always larger than the corresponding critical z value, due to the fact that central t distributions have thicker tails than the z distribution.

Standardized confidence intervals Unfortunately the principles used to construct exact raw CIs cannot be used to construct exact standardized CIs when the population standard deviation is unknown, because we cannot divide the raw CI limits by the population standard deviation. We can, of course, divide the raw limits by the sample standard deviation, thereby producing an approximation to an exact standardized CI. (It is worth repeating here that an 'exact' CI is one produced by a procedure controlling the noncoverage error rate exactly over an indefinitely large series of replications of the experiment when the relevant assumptions are satisfied. An inference from an exact interval is still subject to error.)

In this particular case the sample standard deviation is $\hat{\sigma} = 15.32$, slightly larger than the population standard deviation of $\sigma = 15.00$. Dividing the relevant raw statistics by 15.32 produces statistics scaled in sample standard deviation units: $M_1 - M_2 = 0.44\,\hat{\sigma}$ and the 95% CI is $(-0.20\,\hat{\sigma},\ 1.08\,\hat{\sigma})$. If we were to interpret –0.20 and 1.08 as the limits of a standardized CI, we would in effect be inferring that $\mu_1 - \mu_2 \in (-0.20\sigma, 1.08\sigma)$, thereby ignoring the distinction between the sample standard deviation and the population standard deviation. While we cannot claim that $(-0.20, 1.08)$ is an exact standardized CI, we can treat it as an approximate standardized CI. In this particular case the approximation is a good one: it turns out that the exact standardized interval produced by the noncentral t procedure developed by Steiger and Fouladi (1997) is (–0.19, 1.07). In most cases where the total sample size is similar to or larger than that in the example (N = 40), interpretations of effect size inferences from approximate standardized intervals should be virtually indistinguishable from those derived from exact intervals. In some cases, however, particularly when the sample size is small and the estimated effect size is large, the approximation can be poor.

If zero is inside (outside) an approximate standardized CI, it will also be inside (outside) the corresponding exact standardized interval and the exact raw CI. That is, the three intervals have the same implication for directional inference.

The relatively complex noncentral t CI procedure is described in Appendix C. While this procedure should be preferred to the central t procedure for the construction of t-based standardized CIs, it cannot be used for the construction of standardized CIs in unrestricted ANOVA-model analyses. Many of the analyses recommended in this book make use of test statistics for which noncentral CI methods are not available.[7]

Replicating the experiment: a simulation

The time has come to reveal the source of the data set we have been analysing. The numbers were generated by a computer in a simulation of what might happen if an experiment was replicated 20 times, each replication including a different random sample of 40 subjects from the same population of potential subjects. The difference between population means was set at $\mu_1 - \mu_2 = 7.5$, so that the standardized mean difference is $(\mu_1 - \mu_2)/\sigma = 7.5/15 = 0.5$, a medium effect according to Cohen's guidelines. The data set from Replication 5 in this series of 20 replications was used for the analyses discussed earlier. (Data from any of the other 19 replications could have been used to demonstrate CI procedures. Replication 5 was chosen because that particular data set happens to illustrate certain principles better than some of the other data sets.)

Raw confidence intervals We can see the operation of the most important principles underlying confident inference on mean differences by examining the variation between intervals across replications. The 95% raw CIs are shown in Figure 1.1, with intervals constructed with the z procedure (assuming known population variance) in the upper panel, and intervals constructed with the t procedure in the lower panel. The most striking feature of these graphs is their similarity – estimating the standard error from the data does not have a substantial impact on the outcomes of the analysis when the relevant t distribution has 38 degrees of freedom. The most important difference between the graphs is the absence of variability in the widths of the intervals produced by the z procedure; the width of all such intervals is 18.59, while the width of the t intervals varies between 16.07 and 22.88.

The broken line shows the value of the population mean difference, so any CI covering this line produces a correct inference, and any interval not covering the line produces a noncoverage error. Both procedures produce a noncoverage error from Experiment 18, and the t procedure also produces a noncoverage error from Experiment 7. We can be confident from statistical theory that if the number of replications in this simulation had been increased from 20 to (say) 10,000, then the percentage of replications with noncoverage errors would be extremely close to 5 for both the z and t procedures. (As Experiment 7 shows, this does not mean that both procedures always produce the same inference.)

Both procedures produce an interval containing only positive values (thereby justifying the correct directional inference $\mu_1 > \mu_2$) from 6 of the 20 replications. (Replications 8 and 9 produce a correct directional inference from

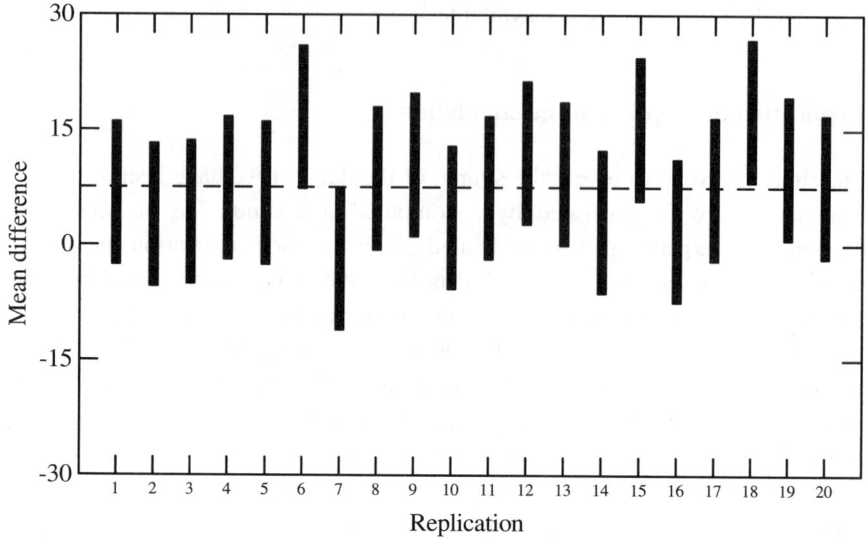

(a) Confidence intervals constructed with the z procedure

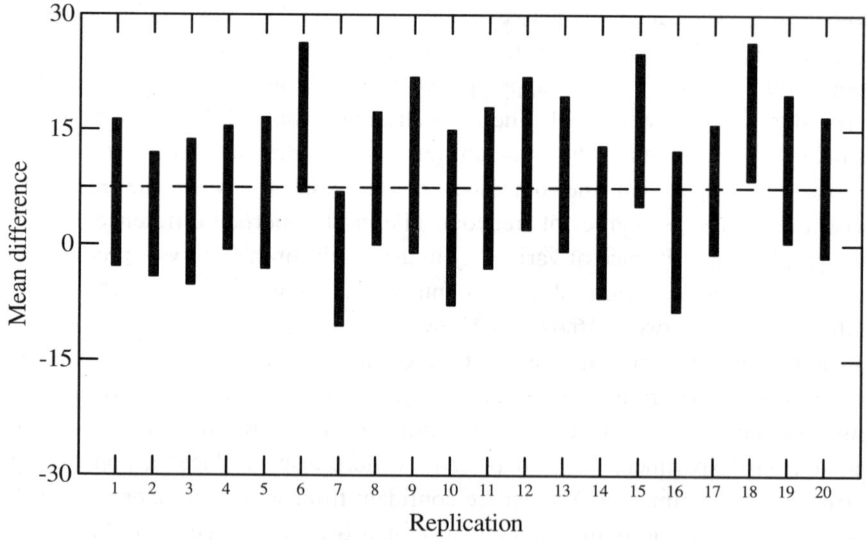

(b) Confidence intervals constructed with the t procedure

Figure 1.1 *Raw 95% confidence intervals on a difference between two means from 20 replications of one experiment*

one but not both of the two procedures.) This result is consistent with expectations from a statistical power analysis, which shows that the power of a two-tailed t test in this context is .34, while the power of a two-tailed z test is .35. Neither procedure produces a Type III error (a directional inference in the wrong direction). Note that a Type I error is not possible here, because the null hypothesis is false.

Standardized confidence intervals Standardized CIs are shown in Figure 1.2. The upper panel shows the exact standardized intervals constructed on the assumption that the population variance is known, and the lower panel shows the approximate standardized intervals obtained by dividing the limits of each raw t interval from Figure 1.1(b) by the relevant estimated standard deviation. Again, the two sets of intervals are similar, indicating that the approximation is probably a reasonable one, at least for this particular combination of sample size and effect size. Unlike the exact raw intervals in Figure 1.1(b), the approximate standardized intervals in Figure 1.2(b) do not vary in width: the width of all 20 intervals is 1.28 (estimated) standard deviation units. The exact standardized intervals from the z (known variance) procedure also have constant width, namely 1.24 population standard deviation units. Unlike the raw intervals, the two sets of standardized intervals differ slightly in their midpoints, because the raw midpoints are divided by slightly different quantities (the constant population standard deviation in the case of the z procedure, and the variable sample standard deviation in the case of the t procedure).

If standardized CIs are to be interpreted in accordance with Cohen's effect size guidelines, then a t-based approximate standardized interval emerging from a given experiment in this series is likely to produce much the same interpretation as a z-based exact interval. Consider, for example, the results from Replication 9, which produced the largest estimated standard deviation (17.88) and the greatest discrepancy (0.11) between the two estimates of the standardized difference between population means. Furthermore, this was the only replication producing a directional inference (a statistically significant difference) from the z procedure but not the t procedure. The z-based interval is (0.08, 1.32), while the t-based interval is (–0.06, 1.22). Both intervals assert that μ_1 is not nontrivially smaller than μ_2, and that μ_1 is practically equivalent to or greater than μ_2. Both intervals fail to provide a precise estimate of the magnitude of the difference between means, excluding only nontrivial negative and very large positive differences.

The sample size ($n = 20$, $N = 40$) used in this simulation is not small, relative to the sample sizes often used in experimental psychology. Two aspects of the simulation data would probably surprise many researchers, particularly those who expect relatively small samples to provide reasonably precise estimates of parameters of interest. First, the width of the standardized CIs (1.28 for t-based

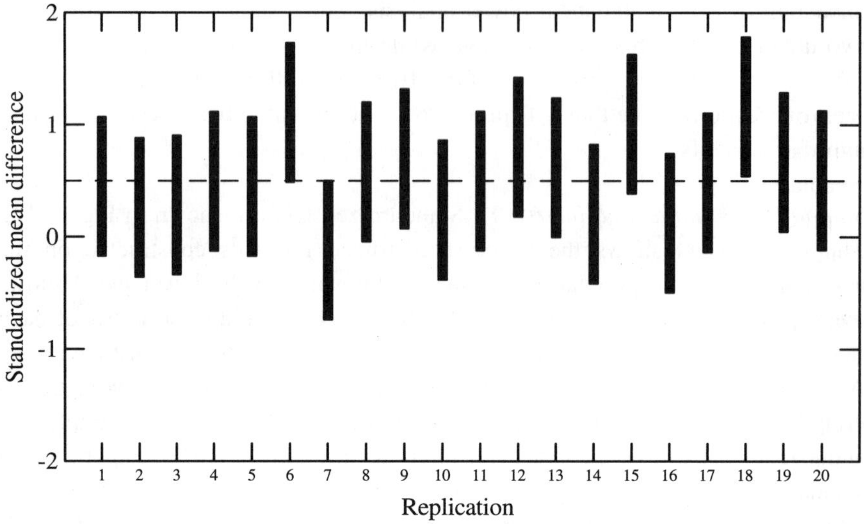

(a) Exact confidence intervals constructed with the z procedure

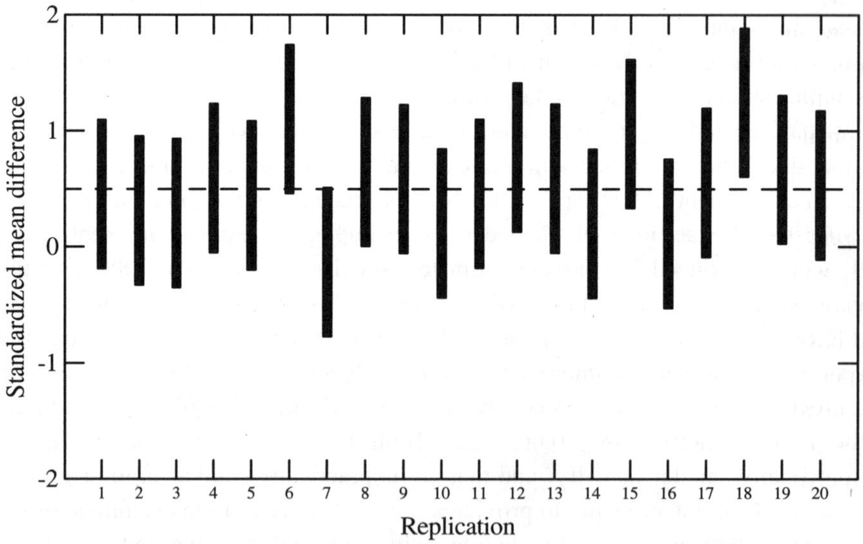

(b) Approximate confidence intervals constructed with the t procedure

Figure 1.2 *Standardized 95% confidence intervals on a difference between two means from 20 replications of one experiment*

intervals) implies that a substantial range of possibly important differences between population means must be included in any given interval. An interval with a midpoint of zero, for example, also includes small and medium positive differences, as well as small and medium negative differences. If this degree of (lack of) precision is unacceptable in a particular context, then the problem is with the sample size, not with the method used to construct the CI. Indeed, an advantage of the CI approach is that it is a relatively simple matter to estimate the precision of estimation (standardized CI width) before the experiment is run. In any two-group experiment with 20 subjects per group, the width of an exact 95% standardized CI produced by the z method will be exactly 1.24, and the width of an approximate 95% interval produced by the t method will be exactly 1.28. If such an interval is deemed to be unacceptably imprecise, it is not difficult, as we will see in Chapter 3, to determine the sample size required to produce a standardized interval of any desired width. Those who are surprised by the width of the intervals in this simulation would probably also be surprised by the magnitude of the variability across replications in the locations (midpoints) of the intervals. The standard deviation of (raw) interval midpoints across an indefinitely large number of replications is simply the standard error used to construct the z-based intervals. (For raw z-based intervals, this figure is 4.74; the standard deviation of the midpoints of the sample of 20 such intervals shown in Figure 1.1(a) is 4.89.) It follows that an experimenter who has access only to data from a single replication is nevertheless able to estimate the variability across replications in CI midpoints from the same statistic (the estimated standard error, which is 4.84 in Replication 5) used to construct the t-based interval emerging from that single replication.

Implications for directional inference Because confident direction inference ($\mu_1 > \mu_2$ or $\mu_1 < \mu_2$) is the highest level of statistical inference aspired to by many researchers, it is of some interest to see what inferences at this level would be possible from the 20 replications in the simulation. Before running such an experiment, a researcher planning to carry out a .05 level two-tailed t test (equivalent to using a 95% CI for directional inference only) would know that if $\mu_1 - \mu_2 = 0$, then the probability of no inference is .95, while the probability of a (necessarily incorrect) directional inference is .05. These probabilities do not apply to the replications in the simulation, because, unknown to the experimenter, the difference between μ_1 and μ_2 is greater than zero by 0.5 standard deviation units. The probability of a correct directional inference ($\mu_1 > \mu_2$) is only .34, while the probability of no directional inference is .66. The probability of a Type III error (getting the direction wrong) in this case is too small to worry about. After the experiment is run, the experimenter knows that in that particular sample from a population of potential replications of the experiment, the test outcome either does or does not justify a confident directional inference. If the experiment happens to be Replication 5 from the simulation, then it turns out that no inference can be justified by the test. (The probability under the null hypothesis of a t at least as large as the obtained value of 1.40 is $p = .17$.) This 'nonsignificant' (no inference) outcome is, needless to say, absolutely uninformative, not because there is anything wrong with the test procedure, but simply because at this level of inference much of the information in a potentially informative 95% CI is ignored. As we saw earlier, the t-based standardized CI (–0.20, 1.08) excludes the possibility of a confident direction inference, but it does support the inference that μ_2 is not nontrivially larger than μ_1, among other things. Unlike the test, the interval also shows that the experiment is not sufficiently sensitive to permit a precise estimate of the magnitude of the effect.

The subjectivist critique of confidence interval inference

CI inference is part of the *classical* or *frequentist* approach to statistical inference, which treats parameters such as population means as fixed and statistics such as sample means as variable. As a consequence, probability statements refer to statistics rather than parameters. Thus, while it may be possible to justify a statement like: 'The probability that a CI constructed in a particular way will cover the parameter is .95', it is considered improper to state that 'the probability that the parameter is covered by this CI is .95'.

CIs are often misinterpreted because the ordinary-language meaning of the word 'probability' is more closely related to the interpretation of that term in the *subjectivist* approach to statistical inference than it is to the interpretation of the same term in the classical approach to inference. The subjectivist (or Bayesian) approach treats parameter values as values of random variables, thereby making it possible to define probability distributions referring to parameters like population means or differences between population means.

The Bayesian inferential framework requires the experimenter to specify a *prior* probability distribution of the parameter of interest. The prior probability distribution is generally interpreted as a distribution of *subjective* probabilities reflecting the researcher's beliefs about the relative credibility of various parameter values, prior to seeing the data. Given this prior distribution and the data (together with an additional probability distribution referring to data rather than parameters), a Bayesian analysis produces a *posterior* distribution of the parameter (a revised distribution taking the data into account). The posterior distribution can be used to construct an interval (usually called a credible interval) for which an interpretation like 'the probability that the parameter lies in this interval is .95' is appropriate. For further details on Bayesian inferential procedures, see Pruzek (1997).

From a Bayesian perspective the classical significance testing and CI approaches are flawed, because they do not allow for probability distributions on parameters (either prior or posterior), and therefore cannot justify the kinds of inferential statements researchers would like to make.

Bayesian inference is not without its critics (see, for example, Oakes, 1986). The most obvious target for criticism is the role played in Bayesian inference by the prior probability distribution, allowing the beliefs and prejudices of an individual researcher to influence the interpretation of the data. This criticism can be dealt with in a number of ways, one of which is to use an 'uninformative' prior distribution, thereby ensuring that prior beliefs have no influence on the outcome of the analysis. It turns out that the use of uninformative prior distributions in a Bayesian analysis produces credible intervals whose limits are similar (if not identical) to those of CIs from a classical analysis (Pruzek, 1997). It would appear, then, that the consequences of common misinterpretations of CIs are less profound in practice than they might appear in theory.

It would be a mistake, however, to ignore the implications of the Bayesian approach to inference, because it does help to make explicit some of the limitations of classical methods of analysis. If an experiment is one of a series, each of which adds something to an existing body of knowledge, a Bayesian analysis can, at least in principle, take that knowledge into account in the specification of the prior distribution. Inferences from a classical analysis, on the other hand, can be influenced only by the data in the current experiment. In a discipline such as psychology where statistical power is typically low (Sedlmeier and Gigerenzer, 1989), classical analyses at the level of individual experiments can be expected to produce imprecise inferences, relative to those sometimes possible from meta-analyses of sets of similar experiments (Schmidt, 1996), or, in a Bayesian framework, analyses of individual experiments that take prior research into account. It does not follow, however, that the rate of incorrect inferences from CIs is likely to be higher than the nominal error rate.

Further reading

Cumming and Finch (2001) and Smithson (2003) provide extensive discussions of CI procedures based on both central and noncentral t distributions. If you feel the need to consolidate your understanding of CI inference before proceeding to Chapter 2, these are good places to start.

Hsu (1996) discusses some of the most important ideas introduced in this chapter (particularly levels of inference and practical equivalence inference). Hsu's treatment does, however, assume a greater degree of mathematical sophistication than is assumed here. For examples of practical equivalence inference in psychology, see Rogers, Howard and Vessey (1993). Cohen (1988),

Richardson (1996) and Rosenthal (1994) provide discussions of effect size measures.

If you would like to become acquainted with the recent history of the debate in psychology concerning the relative value of CIs and significance tests, you will find it worthwhile to consult Cohen (1994), Nickerson (2000), Schmidt (1996), or some of the relevant chapters in Harlow, Mulaik and Steiger (1997). For an indication of the approach to this issue recently adopted by the American Psychological Association, see Wilkinson and the Task Force on Statistical Inference (1999) and the 5th edition of the APA Publication Manual (American Psychological Association, 2001).

Oakes (1986) provides a very readable book-length discussion and critique of a number of approaches to statistical inference, including classical and Bayesian approaches.

Questions and exercises

At the end of each chapter you will find a set of questions and exercises designed to test your understanding of the material in the chapter, and, particularly in subsequent chapters, to provide you with opportunities to practise carrying out relevant analyses. You can check on your answers by consulting Appendix E.

1. In a study designed to investigate the effect of practice on performance on an aptitude test, participants are randomly assigned to one of two experimental conditions. Those in the first (treatment) condition are given practice on items similar to those in the aptitude test, while those in the second (control) condition spend the same amount of time answering questions in an interest inventory. The experimenters are primarily interested in knowing whether the magnitude of the practice effect is large enough to justify changes in a selection procedure that makes use of the test. A mean practice effect of 3 (items correct) is regarded as the smallest nontrivial effect.

The inference about the practice effect is to be based on a 95% CI on $\mu_T - \mu_C$.

What conclusion (if any), at each of the three levels of confident inference discussed in this chapter (interval, direction and inequality inference), would follow from each of the following CIs:

(a) $\mu_T - \mu_C \in (6.5, 8.7)$

(b) $\mu_T - \mu_C \in (0.9, 16.8)$

(c) $\mu_T - \mu_C \in (-0.6, 1.6)$

(d) $\mu_T - \mu_C \in (-7.4, 8.5)$?

2. Assume that the within-condition standard deviation for the experiment in Question 1 is known (independently of the data) to be $\sigma = 8.2$. Given the raw CIs in Question 1, construct and interpret standardized CIs.

3. Comment on the relative precision of the CIs you constructed in your answer to Question 2. Without doing any additional calculations, comment also on sample sizes in the four cases.

4. What types of inferential error (if any) follow at each level of confident inference (CI, confident direction, confident inequality) from the raw CIs produced by the *t* procedure from Replication 1 and Replication 18 of the simulated experiment discussed on page 17? To answer this question you will need to consult Figure 1.1(b).

5. Suppose that you were an experimenter running the simulated experiment, and you obtained the same data as that produced by Replication 18. Could you answer Question 4 if it referred to your data? If not, why not?

Notes

1. A fixed-effects ANOVA model (the type of model usually implied when the term ANOVA is used without qualification) is appropriate for the analysis of randomized experiments where the various experimental conditions (treatments) are selected by the experimenter. A *random-effects* ANOVA model, otherwise known as a variance components model, is appropriate when treatments are randomly sampled from a population of potential treatments. See Bird (2002) or Smithson (2003) for brief discussions of CI inference on parameters of random-effects models.

2. The term *replication* is used in at least three different senses by statisticians and experimenters. The term is used in this book to refer to a repetition of an experiment that makes use of a different random sample of subjects from the same population, but is otherwise identical to the original experiment (or another replication).

3. The usual derivation of the standard error used in a two-group *t* test (or CI) assumes that the subjects assigned to each treatment are randomly sampled from a population of infinite size. In practice, the 'sample' of $N = 2n$ subjects in a two-group experiment is usually a convenience sample, not a random sample from any population. Given random assignment from a convenience sample, the same standard error can be derived by replacing the random sampling assumption with the assumption that the size of the treatment effect does not vary across subjects (Reichardt and Gollub, 1999). If the size of the treatment effect does vary across subjects, the standard error used by the *t*-test procedure is likely to be too large, so that the inferences from the procedure are valid but conservative, with too few Type I errors from tests and too few noncoverage errors from CIs. Given random assignment from a convenience sample, the parameter $\mu_1 - \mu_2$ refers to the 'population' of all subjects in the experiment. Of course, random sampling

has one important advantage over random assignment: it provides a justification for a generalization beyond the $N = 2n$ subjects in the experiment to the population from which they were sampled. In the terminology of Cook and Campbell (1979), random assignment provides a basis for claims of internal validity, while random sampling provides a basis for claims of both internal and external validity.

4. Glass (1976) suggested that the standardized effect size should be expressed in units of variability in the control population. For various reasons, including the fact that the designation of one treatment as the 'control' is often arbitrary, it has become standard practice to assume that both treatment populations have the same standard deviation, and to use this common standard deviation as the unit of measurement when defining a standardized effect size.

5. It is possible to justify confident practical equivalent inference by carrying out two nonstandard tests, one allowing for the possibility of the inference $\mu_1 - \mu_2 < \tau$, the other allowing for the possibility of the inference $\mu_1 - \mu_2 > -\tau$ (Rogers, Howard and Vessey, 1993). The CI approach, however, is simpler and more informative.

6. Following Lockhart (1998) and Smithson (2000), the symbol w is used to refer to the half-width (rather than the width) of a CI. The width of a CI is therefore $2w$.

7. Many ANOVA-model analyses make use of an F test statistic, critical values of which can be used to construct CIs on various monotonic functions of the noncentrality parameter of a noncentral F distribution. (See Appendix C for details.) In most cases it is not possible to transform a CI on this noncentrality parameter into a CI on contrasts (generalized comparisons). A number of ANOVA-model analyses use test statistics for which noncentral interval estimation procedures have not been developed.

2 One-way Analysis of Variance

The ANOVA model

In this chapter we consider the simplest version of the ANOVA *model* and examine various procedures for making inferences about the parameters of this model. We can think of an ANOVA model as a theory of the data produced by an experiment. The single-factor fixed-effects ANOVA model, appropriate for simple randomized experiments with J groups ($J \geq 2$), can be written as

$$Y_{ij} = \mu + \alpha_j + \varepsilon_{ij} \tag{2.1a}$$

or

$$E(Y_{ij}) = \mu + \alpha_j, \tag{2.1b}$$

where μ is a constant,

α_j is an *effect parameter* referring to treatment j

and ε_{ij} is the value of subject i in group j on an *error* variable ε with a mean of zero within treatment population j.

The right hand side of (2.1a) contains three terms (two parameters and the value of a random variable) that account for, or (in a loose sense) explain, the score on the dependent variable of subject i in group j. None of the terms on the right hand side is at any stage known to the experimenter, who must therefore make inferences about them from an appropriate analysis. The error component ε_{ij} is the only term with an i subscript, so it is the only component of the model that can account for individual differences in dependent variable scores between different subjects in the same group. According to (2.1b), the systematic component of the scores of subjects in group j is the sum of two parameters: the constant μ and the effect parameter α_j. If the effect of the first treatment on dependent variable scores is different from the effect of the second treatment, then that difference must be reflected in the difference between α_1 and α_2.

Inferences about effect parameters depend on the following assumptions about error distributions:[1]

- for any pair of Y values in the experiment, the associated ε values are statistically independent;
- the variance of ε_j is the same in all J treatment populations ($\sigma^2_{\varepsilon_j} = \sigma^2_{\varepsilon}$ for all j);

- within each treatment population, ε values are normally distributed.

ANOVA as an overparameterized model The number of parameters in the model is $J + 1$ (J effect parameters plus the constant μ), one more than the number of treatments. As a consequence, individual parameters cannot be uniquely defined in terms of the J population means μ_j. If $J = 2$ and we somehow knew that $E(Y_{i1}) = \mu + \alpha_1 = 16$ and $E(Y_{i2}) = \mu + \alpha_2 = 10$, then we could not determine the values of μ, α_1 or α_2. We could, however, determine the values of the *difference* between the effect parameters:

$$\alpha_1 - \alpha_2 = (\mu + \alpha_1) - (\mu + \alpha_2) = 16 - 10 = 6.$$

A difference between effect parameters is an example of an *estimable function* of the parameters of an overparameterized model. Estimable functions of the parameters of (2.1) are uniquely defined, even if the parameters themselves are not. In this book we will be concerned primarily with CI inference on estimable functions of the parameters of various ANOVA models. In particular, we will be concerned primarily with interval estimates of the values of *contrasts* (generalized comparisons) on effect parameters. It turns out that for most of the ANOVA models we will be concerned with, contrasts on effect parameters can also be expressed as contrasts on means, so that we will be able to base much of our discussion on the latter, thereby avoiding unnecessary complexity.

For some purposes it is convenient to be able to work with a restricted version of the ANOVA model where *constraints* are imposed on parameters. The standard constraint is $\sum \alpha_j = 0$: that is, the effect parameters must sum to zero. This zero-sum constraint implies that $\alpha_J = -(\alpha_1 + \alpha_2 + \cdots + \alpha_{J-1})$, so that the last effect parameter is redundant, given the rest. If the α_J parameter in the model is replaced with $-(\alpha_1 + \alpha_2 + \cdots + \alpha_{J-1})$, then the revised version of the model contains only J *nonredundant* parameters: μ and all of the α_j parameters except the last. Given the constraint $\sum \alpha_j = 0$, it is possible to express ANOVA-model parameters in terms of the parameters of the *means model*

$$Y_{ij} = \mu_j + \varepsilon_{ij} \tag{2.2}$$

as follows:

$$\mu = \frac{\sum_j \mu_j}{J}$$

and $\alpha_j = \mu_j - \mu.$

The zero-sum constraint also implies that if the effect parameters are homogeneous (that is, if $\alpha_1 = \alpha_2 = \cdots = \alpha_J$), then they must all be equal to zero. Unless otherwise stated, all references to ANOVA models in this book will be to models with zero-sum constraints.

Effect size

Before proceeding further we should consider the impact of the zero-sum constraint on the relationship between effect parameters as defined by the ANOVA model and *effect size* as that term is usually understood. Effect size is usually defined as a difference between two means, such as the difference between a treatment mean μ_1 and a control mean μ_2. In a two-group experiment, the ANOVA model (with the constraint $\sum \alpha_j = 0$) would define two effect parameters

$$\alpha_1 = \mu_1 - \mu = \mu_1 - \frac{\mu_1 + \mu_2}{2} = \frac{\mu_1 - \mu_2}{2}$$

and $$\alpha_2 = \mu_2 - \mu = \mu_2 - \frac{\mu_1 + \mu_2}{2} = \frac{\mu_2 - \mu_1}{2}.$$

The first effect parameter (α_1) is half the size of the conventionally defined effect ($\mu_1 - \mu_2$), while the second is $-\alpha_1$ because of the constraint that the two must sum to zero. The conventionally defined effect size is a *difference* between ANOVA-model effect parameters: $\mu_1 - \mu_2 = \alpha_1 - \alpha_2$, an estimable function of those parameters that is not influenced by the zero-sum constraint. The fact that $\mu_1 - \mu_2$ is twice as large as the effect parameter α_1 is a consequence of the arbitrary zero-sum constraint; if this constraint were to be replaced with the (equally arbitrary) constraint $\alpha_J = 0$ (perhaps a more attractive constraint for experiments where group J is a control group), then the effect parameter α_1 would be equal to the conventionally defined effect size parameter. Whatever may be said about the influence of constraints on the meaning of effect parameters, however, they have no influence on the meaning of contrasts on effect parameters: any comparison between two effect parameters is always identical to the comparison between the corresponding means, whatever the constraint on effect parameters (if any) may be.

In a two-group experiment the standardized difference between effect parameters $(\alpha_1 - \alpha_2)/\sigma_\varepsilon$ is equivalent to Cohen's *d* (Cohen, 1969).

Cohen's f: a global standardized effect size index Given the zero-sum constraint on effect parameters, the root mean square (RMS) or 'standard deviation' of those parameters

$$\sigma_\alpha = \sqrt{\frac{\sum_j \alpha_j^2}{J}} \tag{2.3}$$

can be interpreted as an overall measure of the magnitude of effects in dependent variable units.[2] Cohen (1969) proposed that the standard deviation of standardized effect parameters

$$f = \frac{\sigma_\alpha}{\sigma_\varepsilon} = \sqrt{\frac{\sum_j \alpha_j^2}{J\sigma_\varepsilon^2}} \tag{2.4}$$

should be used as a global standardized effect size index. Cohen suggested that f values of 0.1, 0.25 and 0.4 could be regarded as small, medium and large effects. These guidelines are consistent with his guidelines for the evaluation of d values in the two-group case. When $J = 2$, $f = |d|/2$.

The f index reflects the overall degree of heterogeneity among the effect parameters (and population means), but it does not provide information about the *pattern* of differences among those parameters when $J > 2$. It is not surprising, therefore, that Cohen's d (which refers to a single difference between two means) has been much more popular than Cohen's f as an effect size index.

An informative analysis of a fixed-effects design with more than two groups must be based on more than one function of the effect parameters defined by the ANOVA model. Informative analyses are usually based on a set of *contrasts* on effect parameters (or population means).

The ANOVA partition of variation

The standard ANOVA *procedure* (as distinct from the ANOVA *model*) begins with a partition of the variation of dependent variable scores into two components, one reflecting variation *within* groups, the other reflecting variation *between* groups. The ANOVA partition of *deviation scores* is

$$Y_{ij} - M = (Y_{ij} - M_j) + (M_j - M) \tag{2.5}$$

where $M = \sum_j \sum_i Y_{ij} \big/ N$

and the partition of *sums of squares* of deviation scores is

$$\sum_j \sum_i (Y_{ij} - M)^2 = \sum_j \sum_i (Y_{ij} - M_j)^2 + \sum_j n_j (M_j - M)^2 \tag{2.6}$$

Equation (2.5) states that deviations of observed dependent variable scores from the overall mean $(Y_{ij} - M)$ can be partitioned into two additive components: one $(Y_{ij} - M_j)$ providing information about variation between individual scores within groups, another $(M_j - M)$ providing information about variation between group means. Equation (2.6) states that the sums of squares of these components are additive. We will illustrate this partition with a small fictitious data set. Dependent variable scores Y_{ij}, group means M_j and the overall mean M follow.

	Group 1	Group 2	Group 3
	107	101	60
	125	105	86
	126	128	91
	131	120	80
M_j	122.25	113.50	79.25
		$M = 105.00$	

Total variation If we square each of the deviation scores on the left hand side of (2.5) and sum the squared values, we obtain the *total sum of squares*:

$$SS_T = \sum_j \sum_i (Y_{ij} - M)^2 = (107-105)^2 + (125-105)^2 + \cdots + (80-105)^2 = 5498.$$

The total sum of squares quantifies the amount of variation in the data, ignoring distinctions between groups.

Variation within groups The *sum of squares within groups* is

$$SS_W = \sum_j \sum_i (Y_{ij} - M_j)^2$$
$$= (107-122.25)^2 + (125-122.25)^2 + \cdots + (80-79.25)^2 = 1366.50.$$

Within-group deviation scores necessarily sum to zero within each group, so there are $J = 3$ constraints on the set of $N = 12$ $(Y_{ij} - M_j)$ values. In any experiment with J groups there are $\nu_W = (N - J)$ degrees of freedom for variation within groups.

The *mean square within groups* is obtained by dividing SS_W by ν_W:

$$MS_W = \frac{SS_W}{\nu_W} = \frac{1{,}366.5}{9} = 151.833.$$

Given the ANOVA-model assumptions about error components and their distributions, MS_W is an unbiased estimator of the error variance σ_ε^2, whatever the magnitude of the effect parameters. For this reason, the within-groups mean square is usually referred to as *mean square error* (MS_E).

Variation between groups The between-group deviation scores $(M_j - M)$ are constant within each group, but they vary across groups. The *sum of squares between groups* is

$$SS_B = \sum_j n_j (M_j - M)^2 = 4\left[(122.5-105)^2 + (113.50-105)^2 + (79.25-105)^2\right]$$
$$= 4\left[(17.25)^2 + (8.50)^2 + (-25.75)^2\right] = 4131.50.$$

The three deviation means (weighted by sample sizes when sample sizes are unequal) necessarily sum to zero:

$$\sum_j n_j(M_j - M) = 4[17.25 + 8.50 + (-25.75)]$$
$$= 69 + 34 + (-103) = 0.$$

This constraint implies that there are only $(J-1)$ independent pieces of information in a set of J deviation means. The number of *degrees of freedom* for variation between groups is $\nu_B = (J-1)$. The *mean square between groups* is obtained by dividing SS_B by ν_B:

$$MS_B = \frac{SS_B}{\nu_B} = \frac{4{,}131.50}{2} = 2065.75.$$

Given the ANOVA model with the zero-sum constraint $(\sum \alpha_j = 0)$, the expected value of MS_B for an equal-n experiment is

$$\mathrm{E}(MS_B) = \sigma_\varepsilon^2 + \frac{n\sum_j \alpha_j^2}{J-1},$$

and the *variance ratio* MS_B / MS_E is distributed as a *noncentral F distribution* with degrees of freedom parameters $\nu_1 = \nu_B = (J-1)$ and $\nu_2 = \nu_E = (N-J)$, and *noncentrality parameter*

$$\delta_F = \frac{n\sum_j \alpha_j^2}{\sigma_\varepsilon^2}. \qquad (2.7)$$

(The corresponding expressions for unequal-n experiments are relatively complex and not particularly enlightening.)

Heterogeneity inference

If the effect parameters are homogeneous (and therefore equal to zero), then $\delta_F = 0$, $\mathrm{E}(MS_B) = \mathrm{E}(MS_E) = \sigma_\varepsilon^2$ and the variance ratio (usually called the ANOVA F statistic) is distributed as the *central F* distribution $F_{J-1,N-J}$. The ANOVA F test rejects the homogeneity hypothesis

$$H_0: \alpha_1 = \alpha_2 = \cdots = \alpha_J = 0$$

if the obtained ANOVA F statistic is greater than $F_{\alpha;J-1,N-J}$, the upper $100(1-\alpha)$th percentile point of the relevant central F distribution. A statistically significant F value justifies the inference that effect parameters are heterogeneous (that is, they are not all equal to each other or to zero). The F test provides no information about the degree of heterogeneity among the effect parameters (or population means).

In an analysis based on CIs it is possible to construct a CI on Cohen's f that implies the outcome of the F test. The f parameter may be expressed in terms of the noncentrality parameter δ as

$$f = \sqrt{\frac{\delta}{N}}, \tag{2.8}$$

so that a recently developed method of constructing exact *noncentral* CIs on δ (Steiger and Fouladi, 1997) can also produce exact CIs on f. (See Appendix C for an account of noncentral CI procedures.) Consider a randomized experiment with $J = 4$ groups, $n = 20$ subjects per group ($N = 80$ subjects in all), and an ANOVA F statistic of $F_{3,76} = 5.914$. (This is the F statistic obtained from the data set we will use to illustrate CI procedures for contrasts.) The 90% CI inference on f is $f \in (0.237, 0.641)$. Cohen's guidelines for the interpretation of f suggest that the overall effect size (that is, the degree of heterogeneity of effect parameters) is somewhere between medium and very large. Because the CI implies that $f > 0$, it also implies that the homogeneity hypothesis is false. The α-level ANOVA F test rejects the homogeneity hypothesis if (and only if) a $100(1 - 2\alpha)\%$ CI on f (a 90% CI if $\alpha = .05$) has a lower limit greater than zero.[3] The F test is redundant, given the CI on f.

Practical equivalence inference If the upper limit of the CI on f is sufficiently small to justify the inference that there is only a trivial degree of heterogeneity among the effect parameters, then the experimenter can conclude that $\alpha_1 \approx \alpha_2 \approx \cdots \approx \alpha_J \approx 0$ (where $\approx$ means 'is practically equivalent to') so that $\mu_1 \approx \mu_2 \approx \cdots \approx \mu_J$. This kind of practical equivalence inference could be used to justify a decision to drop the effects parameters from the ANOVA model, and to model the data instead with the simpler no-effects model $Y_{ij} = \mu + \varepsilon_{ij}$. If a *model comparison* (in this case a comparison between a model including effect parameters and a no-effects model) can be used to justify a decision to adopt the simpler of two competing models, then the analysis is simplified considerably. In principle, this kind of model comparison can be very useful in analyses of complex factorial designs (such as those we will encounter in Chapter 5) because the adoption of a relatively simple model can substantially reduce the complexity of the analysis of *contrasts* on model parameters.

In practice, very large sample sizes are required to allow for the possibility that a CI-based model comparison might produce evidence of a trivial degree of heterogeneity in a set of effect parameters. Suppose, for example, that a four-group experiment with $N = 80$ had produced an F statistic of $F_{3,76} = 1.027$, the expected value when all effect parameters are zero. The 90% CI on f for this outcome is $f \in (0, 0.331)$. The lower limit implies that a .05-level F test would not reject the homogeneity hypothesis (the p value associated with this test is .385), and the values near the lower end of the CI are compatible with the possibility that all of the effect parameters are zero or trivially small. The CI also includes medium and medium–large values of *f*, however, so it does not justify the inference $\alpha_1 \approx \alpha_2 \approx \alpha_3 \approx \alpha_4 \approx 0$, even though the obtained *F* is as small as could reasonably be expected if all population means were equal.

CI inference on the degree of heterogeneity among effect parameters (or population means) is usually of very limited value. Even if it suggests that some of the ANOVA-model effect parameters must be substantial, an inference on *f* provides no information about the nature of these effects. Nevertheless, CI inference on *f* is more informative than the kind of inference justified by the traditional ANOVA *F* test. As will see, even the meagre analytic yield from a CI inference on *f* is often redundant (and sometimes irrelevant) in the context of an analysis based on contrasts.

Contrasts

In the context of a single-factor ANOVA-model analysis, a contrast is a linear combination of effect parameters where the coefficients in the linear combination sum to zero. Analyses of data from multiple-group experiments are usually based on multiple contrasts. Let ψ_g be the *g*th contrast in an analysis, defined by a set of *J contrast coefficients* c_{gj}. Then the value or magnitude of the contrast is

$$\psi_g = \sum_j c_{gj}\alpha_j = c_{g1}\alpha_1 + c_{g2}\alpha_2 + \cdots + c_{gJ}\alpha_J \qquad \left(\sum_j c_{gj} = 0\right). \tag{2.9}$$

If $c_{11} = 0.5$, $c_{12} = 0.5$, $c_{13} = -1.0$ and the remaining coefficients c_{1j} are all zero, then the contrast ψ_1 is $0.5\alpha_1 + 0.5\alpha_2 - \alpha_3$, the difference between the average of the first two effect parameters and the third. In the context of an analysis based on the means model, the same set of contrast coefficients defines a linear combination of means rather than effect parameters. The contrast

$$\psi_g = \sum_j c_{gj}\mu_j = c_{g1}\mu_1 + c_{g2}\mu_2 + \cdots + c_{gJ}\mu_J \qquad \left(\sum_j c_{gj} = 0\right) \tag{2.10}$$

is the same contrast as that defined by (2.9) if the coefficients are the same in both cases: the contrast $\psi_1 = 0.5\alpha_1 + 0.5\alpha_2 - \alpha_3$ is also the contrast $\psi_1 = 0.5\mu_1 + 0.5\mu_2 - \mu_3$. When carrying out analyses based on the single-factor ANOVA model, it is usually easier to think of contrasts as linear combinations of means rather than as linear combinations of effect parameters.

The symbol ψ_g is used for two purposes. First, it refers to a particular contrast in an analysis that may include several contrasts: if a second contrast is defined with coefficients $c_{21} = 1$ and $c_{22} = -1$ (with all other coefficients zero), then $\psi_2 = \mu_1 - \mu_2$ is not the same contrast as ψ_1. Second, the magnitude of ψ_g is the *population value* of a particular contrast. Consider an experiment with $J = 3$ conditions (two different treatments for a disorder and a control condition) where the population means are $\mu_1 = 124$, $\mu_2 = 110$ and $\mu_3 = 85$. The population value of the contrast comparing the average of the two treatment means with the control mean is

$$\psi_1 = 0.5\mu_1 + 0.5\mu_2 - \mu_3 = 62 + 55 - 85 = 32,$$

and the population value of the contrast comparing the two treatment means is

$$\psi_2 = \mu_1 - \mu_2 = 124 - 110 = 14.$$

Vector notation It will be convenient to make occasional use of vector notation when referring to individual contrasts. A vector is a column (or row) of symbols or numbers. The set of means defined in the previous paragraph can be written in vector notation as $\boldsymbol{\mu}' = [\mu_1 \;\; \mu_2 \;\; \mu_3] = [\, 124 \;\; 110 \;\; 85 \,]$. We can refer to $\boldsymbol{\mu}'$ as the vector of population means, the prime indicating that we are choosing to write the vector as a row rather than as a column. (The vector $\boldsymbol{\mu}'$ is the *transpose* of $\boldsymbol{\mu}$, a vector containing the same set of means written as a column.) The orientation of a vector can be important for the purpose of carrying out mathematical operations involving vectors, but is usually arbitrary otherwise.

We can define an individual contrast compactly in terms of its *coefficient vector* $\mathbf{c}_g$ (or $\mathbf{c}'_g$). The coefficient vectors for the contrasts ψ_1 and ψ_2 are $\mathbf{c}'_1 = [\, 0.5 \;\; 0.5 \; -1 \,]$ and $\mathbf{c}'_2 = [\, 1 \;\; -1 \;\; 0 \,]$. If you are familiar with matrix algebra you will recognize that $\mathbf{c}'_1 \boldsymbol{\mu}$ is the *inner product* of the contrast coefficient vector $\mathbf{c}'_1$ and the mean vector $\boldsymbol{\mu}$, a compact way of summarizing the *sum of products* of contrast coefficients and population means given by (2.10) as the definition of a contrast on means. That is, $\psi_1 = \mathbf{c}'_1 \boldsymbol{\mu}$ is a compact way of defining the contrast ψ_1. We will occasionally use vector or matrix notation to summarize simple operations involving the calculation of sums of products. In the main, however, we will use vector (or matrix) notation to refer to individual vectors (or matrices), not to provide a basis for the mathematical operations of matrix algebra. Familiarity with matrix algebra will not be assumed.

The scale of contrast coefficients

The magnitude of the population value of a contrast depends on the vector of population means and the *pattern* and *scale* of the contrast coefficients. In the case of ψ_1 with coefficient vector $[\, 0.5 \;\; 0.5 \; -1 \,]$, the pattern of coefficients is identical to that of any other contrast with coefficients proportional to those in $\mathbf{c}'_1$. For example, the coefficient vector $\mathbf{c}'_{1b} = 2\mathbf{c}'_1 = [\, 1 \;\; 1 \; -2 \,]$ has the same pattern as $\mathbf{c}'_1$: the two sets of coefficients are proportional and therefore perfectly correlated. The value of $\psi_{1b} = \mathbf{c}'_{1b} \boldsymbol{\mu}$ must be twice as large as the value of ψ_1, simply because of the difference between $\mathbf{c}_1$ and $\mathbf{c}_{1b}$ in the *scale* (magnitude) of the contrast coefficients. Any interpretation of the magnitude of a contrast must take into account the scaling of the contrast coefficients.

Mean difference contrasts A mean difference contrast compares the (weighted or unweighted) average of a subset of means with the average of a different (nonoverlapping) subset. A vector of contrast coefficients is scaled to define a mean difference contrast if the positive coefficients sum to 1.0; that is, if $\sum c^+ = 1.0$, where c^+ denotes a positive coefficient.[4] The scaling of the coefficients in $\mathbf{c}_1$ satisfies this criterion ($\sum c_1^+ = 0.5 + 0.5 = 1.0$), so ψ_1 is a mean difference contrast. The scaling of coefficients in $\mathbf{c}_{1b}$ does not satisfy the mean difference criterion because $\sum c_{1b}^+ = 2$. The coefficient vector for any contrast with arbitrary scaling (such as ψ_{1b}, the difference between the sum of the treatment means and twice the sum of the control means) can be rescaled to define a mean difference contrast by dividing each coefficient by $\sum c_j^+$.

Following Scheffé (1953), we will refer to a mean difference contrast comparing the *unweighted* average of a subset of *m* means with the unweighted average of a different nonoverlapping subset of *r* means ($m + r \leq J$) as an $\{m, r\}$ contrast. ψ_1 is a $\{2, 1\}$ contrast, and any comparison (such as ψ_1) is a $\{1, 1\}$ contrast. All $\{m, r\}$ contrasts are mean difference contrasts, but mean difference contrasts involving *weighted* averages of subsets of means are not $\{m, r\}$ contrasts.

Mean difference scaling is appropriate for most contrasts in *single-factor* experiments. (In a single-factor experiment we can think of each treatment or experimental condition as a level or value of a single categorical variable or factor. In a multifactor experiment each experimental condition is a combination of levels of two or more categorical variables. We will consider multifactor experiments in Chapter 4.) Mean difference scaling is required if an experimenter wishes to use Cohen's (1969) guidelines to interpret the magnitude of a standardized contrast as a measure of effect size. Suppose, for example, that $\sigma_\varepsilon = 16$, so that the standardized difference between the average of the three treatment means and the control mean is $\psi_2 / \sigma_\varepsilon = 32 / 16 = 0.5$. (This is also the average of the standardized values of the two comparisons comparing one of the two treatment means with the control mean.) While it is reasonable to interpret the standardized difference defined by the $\{2,1\}$ contrast ψ_1 as a medium effect (in Cohen's sense), it would make no sense to interpret $\psi_{1b} / \sigma_\varepsilon = 2\psi_1 / \sigma_\varepsilon = 1.0$ as a large effect.

Mean difference scaling is inappropriate if a contrast is to be interpreted as a difference between two mean differences: a contrast comparing an effect size (defined as the difference between the means of a treatment condition and a control condition) for male subjects with the effect size for female subjects is not a mean difference contrast. Contrasts of this kind, sometimes called *contrasts of contrasts* (Harris, 1994), are rarely used in the analysis of single-factor designs, but are frequently used in analyses of multifactor designs, as we will see in later chapters.

Contrast coefficient scaling and directional inference For some purposes the scale of contrast coefficients is irrelevant. Confident direction and confident inequality inferences do not depend on the scale of contrast coefficients. If (and only if) the population value of ψ_1 is positive (negative), then the population value of any other contrast with proportional coefficients (such as ψ_{1b}) will be positive (negative). Further, any confident inference procedure producing a directional inference on a particular contrast will produce the same directional inference on any other contrast with proportional coefficients. For this reason, contrasts with proportional coefficients are sometimes called *identical* contrasts. It is important to realize, however, that so-called identical contrasts with different coefficient scaling cannot have the same value unless that value is zero. The scale of coefficients cannot be ignored at the CI level of inference.

Contrast statistics

An unbiased estimate of the population value of a contrast can be obtained from data from a randomized experiment by substituting sample means for population means in (2.10):

$$\hat{\psi}_g = \sum_j c_{gj} M_j = c_{g1} M_1 + c_{g2} M_2 + \cdots + c_{gJ} M_J \qquad \left(\sum_j c_{gj} = 0 \right). \qquad (2.11)$$

In vector notation, $\hat{\psi}_g = \mathbf{c}'_g \mathbf{m}$, where $\mathbf{m}$ is the vector of sample means.
The standard error of the contrast sample value $\hat{\psi}_g$ is

$$\sigma_{\hat{\psi}_g} = \sigma_\varepsilon \sqrt{\sum_j \frac{c_{gj}^2}{n_j}} \, . \qquad (2.12)$$

In standard deviation units, the standard error is

$$\frac{\sigma_{\hat{\psi}_g}}{\sigma_\varepsilon} = \sqrt{\sum_j \frac{c_{gj}^2}{n_j}} \, . \qquad (2.13)$$

When sample sizes are equal (an important special case), the standardized standard error of a contrast is

$$\frac{\sigma_{\hat{\psi}_g}}{\sigma_\varepsilon} = \sqrt{\frac{\sum_j c_{gj}^2}{n}} \, . \qquad (2.14)$$

It is clear from (2.14) that, at least when sample sizes are equal, the standard error of a contrast sample value decreases as the sample size increases and as the sum of squares of the contrast coefficients decreases. The influence of $\sum c^2$ on the standard error of an $\{m, r\}$ contrast, which is really an indirect influence of the complexity of the contrast, is worthy of examination. The {2,1} contrast ψ_1 with coefficient vector

$$\mathbf{c}_1' = [\,0.5 \;\; 0.5 \; -1\,] \quad \text{and} \quad \sum_j c_{1j}^2 = (0.5)^2 + (0.5)^2 + (-1)^2 = 1.5$$

has a standardized standard error of $\sqrt{1.5/n}$, whereas the comparison ψ_2 with coefficient vector

$$\mathbf{c}_2' = [\,1 \;\; -1 \;\; 0\,] \quad \text{and} \quad \sum_j c_{2j}^2 = 2.0$$

has a standardized standard error of $\sqrt{2/n}$. It follows that an experiment with n subjects per group will produce a more precise estimate of the value of a {2, 1} contrast than of the value of a comparison. The reason for this difference in precision is that whereas $\hat{\psi}_2$ compares the mean of n subjects in group 1 with the mean of n subjects in group 2, $\hat{\psi}_1$ compares the mean of the $2n$ subjects in groups 1 and 2 (that is, all subjects given a treatment) with the mean of the n subjects in group 3. Thus the *effective sample size* for the complex contrast is larger than that for the comparison.[5]

The standard error of the sample value of a contrast can be estimated by substituting $\sqrt{MS_E}$ for σ_ε in (2.12):

$$\hat{\sigma}_{\hat{\psi}_g} = \sqrt{MS_E \sum_j \frac{c_{gj}^2}{n_j}}\,. \tag{2.15}$$

The sample value and its estimated standard error are the only statistics required for the construction of a CI on the population value of a contrast. A central CI on any contrast can be expressed as follows:

$$\psi_g \in \hat{\psi}_g \pm \text{CC} \times \hat{\sigma}_{\hat{\psi}_g}, \tag{2.16}$$

where CC is a *critical constant* that depends on the type and magnitude of the nominated noncoverage error rate and on whether the contrast is defined independently of the data. For *planned* contrasts (that is, contrasts defined independently of the data, preferably before the experiment is run), the critical constant is usually a critical value of a central t distribution with $N - J$ degrees of freedom. Given the assumptions underlying ANOVA-model analyses, the *per-contrast* noncoverage error rate (PCER) can be controlled (at α) by setting the CC at $t_{\alpha/2;\,N-J}$. (Note that the symbol α is used to denote error rates as well as effect parameters. The intended meaning is usually obvious from the context.) This is but one of a number of CI procedures we will discuss in this chapter, all of which are special case of (2.16). All of the remaining procedures are designed to address *multiplicity* issues associated with consequences of constructing and interpreting CIs on more than one contrast in a single analysis.

Simultaneous inference on multiple contrasts

Multiplicity issues in ANOVA are usually discussed in the context of analyses restricted to comparisons ({1, 1} contrasts), and we will begin our discussion in that context. A popular approach to the analysis of data from an experiment with several groups is to carry out a statistical test on every comparison, then base the interpretation of the data set on directional inferences on those comparisons that turn out to be statistically significant by the test procedure used. If each test controls the Type I error rate at a conventional α level, then when all effect parameters are zero (so that all population means are equal), the probability of at least one Type I error somewhere in the set of inferences must be greater than α. It is generally agreed within some disciplines (such as psychology) that this 'inflation' of the Type I error rate is unacceptable. The standard solution to this multiple inference problem is to define and control a Type I error rate referring to inferences on the *set* of comparisons, rather than allowing an error rate of α for the inference on each comparison. If the error-rate *unit* is the set of inferences, then $(1 - \alpha)$ is usually defined as the probability of making no Type I errors on any comparison in the set, so that α is defined as the probability of one or more errors.

Similar issues arise with the definition of the error-rate unit for CIs on multiple comparisons (or contrasts). If the confidence level is set at $100(1 - \alpha)\%$ for each CI on a set of k multiple comparisons (or contrasts), then the confidence level for the *set* of CIs (defined as the probability that every CI will cover the population value of the relevant contrast) can approach $100(1 - k\alpha)\%$, so that the probability of one or more noncoverage errors can approach $k\alpha$. Multiple CIs constructed in this way are sometimes called *individual* CIs, and the associated noncoverage error rate is called the *per-contrast* error rate (PCER).

Simultaneous CIs (SCIs) control the confidence level for a set (or *family*) of inferences at $100(1 - \alpha)\%$; if 95% SCIs are constructed on several contrasts, then the probability is at least .95 that all of these intervals will cover their respective population contrast values. SCIs are more *conservative* than individual CIs with the same nominal α: they are wider (less precise) and produce fewer noncoverage errors than individual CIs. A CI procedure that produces SCIs controls the *familywise* error rate (FWER), defined as the probability of one or more noncoverage errors in the set of CIs on a number of different contrasts.

Conventions about error-rate units vary across disciplines; familywise error-rate control is generally regarded as appropriate in psychology, but it is usually regarded as unnecessarily conservative in some other disciplines such as epidemiology. We may note in passing that the traditional ANOVA F test is not compatible with individual CIs on contrasts, which can sometimes imply heterogeneity when the F test does not. The F test is, however, compatible with $100(1-\alpha)\%$ SCIs produced by the F-based Scheffé procedure that we will encounter shortly.

Simultaneous confidence interval procedures

Planned contrasts When all of the k contrasts in an analysis are defined independently of the data (preferably before the experiment is run), the FWER can be controlled at α by setting the per-contrast error rate at α/k, thereby setting the per-contrast confidence level at $100(1-\alpha/k)\%$. This adjustment to the α level is known as a Bonferroni adjustment. The *Bonferroni-t* procedure uses a CC of $t_{\alpha/2k;\,N-J}$, which when $k > 1$ is always larger than the unadjusted CC ($t_{\alpha/2;\,N-J}$) used to control the PCER in a planned analysis. SCIs in a Bonferroni-t analysis are constructed from

$$\psi_g \in \hat{\psi}_g \pm t_{\alpha/2k;\,N-J} \times \hat{\sigma}_{\hat{\psi}_g} \qquad (g = 1, 2, \ldots, k). \tag{2.17}$$

For the small data set we have been using to illustrate the ANOVA partition of variation (with $J = 3$, $N = 12$, $MS_E = 151.833$, $\mathbf{m}' = [\,122.25\ \ 113.50\ \ 79.25\,]$), the sample values of the contrasts $\psi_1 = [\,0.5\ \ 0.5\ {-1}\,]\boldsymbol{\mu}$ and $\psi_2 = [\,1\ {-1}\ \ 0\,]\boldsymbol{\mu}$ are $\hat{\psi}_1 = \mathbf{c}_1'\mathbf{m} = 38.625$ and $\hat{\psi}_2 = \mathbf{c}_2'\mathbf{m} = 8.75$, and the estimated standard errors of these sample values [from (2.15)] are $\hat{\sigma}_{\hat{\psi}_1} = 7.5457$ and $\hat{\sigma}_{\hat{\psi}_2} = 8.7130$. The CC for Bonferroni-$t$ SCIs on these two contrasts is $t_{.05/(2\times2);\,9} = 2.6850$. (If required for hand calculations, Bonferroni-t critical values can be obtained from the *PSY Probability Calculator*, which is discussed in Appendix A.)

Given these statistics and the CC, we can construct the SCIs as follows:

$$\begin{aligned}\psi_1 &\in 38.625 \pm 2.685 \times 7.5457\\ &\in (18.365,\ 58.885)\\ \psi_2 &\in 8.750 \pm 2.685 \times 8.713\\ &\in (-14.644,\ 32.144).\end{aligned}$$

Each of these Bonferroni-t SCIs is 18.7% wider than an individual 95% CI on the same contrast, because the Bonferroni-t CC of 2.685 is 18.7% larger than $t_{.05/2;\,9} = 2.262$, the CC for individual CIs. Most of this difference can be attributed to the fact that SCIs are always less precise than individual CIs on the same contrasts, a consequence of the fact that the former control the FWER rather than the PCER. Some of the difference is due to the fact that the FWER produced by the Bonferroni-t procedure is always less than the nominal error rate α. This conservatism (relative to the nominal FWER) is usually trivial if the

k planned contrasts are *linearly independent*, so that none of them can be expressed as a linear combination of the others. The two contrasts in this analysis are linearly independent. The contrasts in the set of $k = 3$ comparisons on three means are not linearly independent, however, because each can be expressed as a linear combination of the other two. For example, $(\mu_1 - \mu_2) = (\mu_1 - \mu_3) - (\mu_2 - \mu_3)$.

Post hoc contrasts An *unrestricted* contrasts analysis can include any contrasts of interest, including *post hoc* contrasts: those that occur to the experimenter after an inspection of the pattern of sample means. Claims about the control over error rates produced by any t-based CI procedure (including the Bonferroni-t procedure) are based on the assumption that a replication of the experiment would include a replication of the choice of contrasts in the analysis, so that the experimenter would have no freedom to vary the choice of contrasts from one replication to the next. The choice of post hoc contrasts in an unrestricted analysis depends on the pattern of sample means in one particular replication, and can therefore vary across replications, thereby invalidating t-based inferential procedures. It is easy to demonstrate by Monte Carlo methods (computer simulations of the outcomes of very large numbers of replications of an experiment) that the FWER produced by the Bonferroni-t procedure can be much greater than α when it is applied to contrasts chosen on a post hoc basis.

The *Scheffé* SCI procedure (Scheffé, 1953) uses a CC derived from the same critical F value used for the ANOVA F test. The Scheffé CC is $\sqrt{\nu_B F_{\alpha;\nu_B,\nu_E}}$, so Scheffé SCIs are constructed from

$$\psi_g \in \hat{\psi}_g \pm \sqrt{\nu_B F_{\alpha;\nu_B,\nu_E}} \times \hat{\sigma}_{\hat{\psi}_g}. \tag{2.18}$$

The Scheffé CC for the current example is $\sqrt{2F_{.05;\,2,\,9}} = 2.9177$, which is 8.7% larger than the Bonferroni-t CC. This means that Scheffé SCIs are less precise than Bonferroni-t SCIs when the analysis is restricted to $k = 2$ planned contrasts. The Scheffé procedure is always *unnecessarily* conservative in analyses restricted to $k \leq (J - 1)$ planned contrasts, because the Bonferroni-t procedure always produces greater precision in these applications. The Bonferroni-t CC increases as k increases, however, and in some planned analyses is larger than the Scheffé CC. For example, in a planned analysis based on the complete set of six $\{m, r\}$ contrasts (three $\{2, 1\}$ contrasts and three comparisons), the Bonferroni-t CC of $t_{.05/(2\times6);\,9} = 3.3642$ would be 15.3% larger than the Scheffé CC of 2.9177, so the Scheffé procedure would be chosen for this application.

The Scheffé procedure produces an FWER of exactly α when the *maximal contrast* (the post hoc contrast with coefficients chosen to maximize the *contrast F* statistic that will discussed later in this chapter) is always included in the set of contrasts to be estimated, and it therefore has no competitors when experimenters wish to control the FWER in completely unrestricted analyses.

Maximal contrasts are almost never of interest to experimenters, however. (The mean difference version of the maximal contrast from the small data set we have been examining is $0.67\mu_1 + 0.33\mu_2 - \mu_3$.)[6] The Scheffé procedure is conservative when applied to the contrasts actually included in analyses.

Although neither the Bonferroni-*t* nor the Scheffé SCI procedure is optimal (in the sense of providing the narrowest possible SCIs in a given application), they are both relatively efficient when used appropriately. The Bonferroni-*t* procedure works well in planned analyses when the number of planned contrasts is not large relative to the number of degrees of freedom between groups. The Scheffé procedure usually provides reasonably efficient post hoc analyses. Various special-purpose SCI procedures can provide a slight increase in precision in particular types of analyses, and we will consider some of these after we have seen how the procedures can be implemented with the *PSY* program.

Heterogeneity inference from computer programs

We will illustrate the CI procedures discussed in this chapter with analyses of data from a hypothetical randomized experiment with $J = 4$ 'treatments' for depression and $N = 80$ subjects who meet the standard criteria for clinical depression ($n = 20$ subjects per condition). One of the treatments is new (NT), one is a standard treatment (ST), one is a minimal-contact treatment (MCT) and the last is a waiting-list control condition (C). The dependent variable is the post-treatment score on a rating scale, a high score indicating freedom from depression. The *PSY* input file *depression.in* (which can be downloaded from http://www.sagepub.co.uk) contains the data.

All of the well-known statistical packages will carry out an ANOVA *F* test. Results are usually displayed in an *ANOVA summary table*, showing sums of squares, degrees of freedom, mean squares and the ANOVA *F* statistic. The Depression data summary table (from *SYSTAT*) follows.

```
                     Analysis of Variance
Source    Sum-of-Squares  df      Mean-Square      F-ratio       P
GROUP        1244.637      3        414.879         5.914      0.001
Error        5331.850     76         70.156
```

The last column (p) shows the probability of obtaining an F statistic at least as large as that obtained if the effect parameters (and population means) are homogeneous. A .05-level ANOVA F test would reject the homogeneity hypothesis in this case because the F statistic is larger than the critical F value ($F_{.05;3,76} = 2.725$) and therefore $p < \alpha$. This test justifies the inference that the effect parameters (and population means) are heterogeneous, but it says nothing about the extent of the heterogeneity.

The *STATISTICA Power Analysis* program can provide an interval estimate of the noncentrality parameter (δ) of the distribution $F_{3,76,\delta}$ that generated the obtained F statistic.[7] Given the F statistic of 5.91367 (and retaining 5 decimal places because this statistic is the basis for additional calculations), the 90% CI produced by *STATISTICA Power Analysis* is $\delta \in (4.48317, 32.90493)$. Two additional calculations (using the relationship $f = \sqrt{\delta/N}$) transform these CI limits on δ into CI limits on f, producing the 90% CI $f \in (0.237, 0.641)$, suggesting that variation between population means is at least moderate and possibly very large, according to Cohen's guidelines (in which f values of 0.1, 0.25 and 0.4 are deemed to indicate small, medium and large effects).

This inference on f is unlikely to be of much interest to the experimenter, because it says nothing about the pattern of differences between means. The traditional ANOVA F test (the outcome of which is implied by the lower limit of the CI on f, provided that α is set at .05 for the test) is even less informative, because it says nothing about the magnitude of effects.

Confidence interval inference on contrasts from computer programs

PSY (Bird, Hadzi-Pavlovic and Isaac 2000) can construct individual or simultaneous central CIs on contrasts referring to parameters of means models or saturated ANOVA models. (The ANOVA model discussed in this chapter and most of the models discussed in later chapters are saturated models.) Although *PSY* was designed primarily for procedures that make use of a CC derived from a critical value of a central t, F, T^2, *GCR* or *SMR* distribution (all of which we will encounter in this book), it can also construct CIs from (2.16) using any CC provided by the data analyst. This means that *PSY* can construct CIs from any central CI procedure. In practice, some *PSY* analyses are very easy to implement, whereas others (particularly those requiring a user-supplied CC) are more difficult for relatively unsophisticated users.

PSY provides (exact) raw CIs and approximate standardized CIs.

SPSS MANOVA can construct central raw CIs (but not standardized CIs) on contrasts referring to the parameters of fixed-effects linear models, including means models, saturated and unsaturated ANOVA models, and more general models such as analysis of covariance (ANCOVA) models. *SPSS MANOVA* is not accessible through menus, and some analyses can be difficult to implement, particularly those involving repeated measures.

SPSS GLM can construct central t-based raw CIs (but not standardized CIs) on a variety of models, and it is often easier to carry out *SPSS GLM* analyses than *SPSS MANOVA* analyses. *SPSS GLM* cannot, however, construct F-based SCIs, or, more generally, SCIs appropriate for unrestricted analyses.

When group sample sizes are equal, *STATISTICA Power Analysis* can construct exact noncentral *t* standardized CIs for planned contrasts.

Although statistical packages will be discussed where appropriate, most of the contrast CI analyses discussed in this book have been implemented by *PSY*, which can be downloaded from http://www.psy.unsw.edu.au/research/psy.htm. If you wish to be able to run *PSY* analyses you should review the first five pages of Appendix A before proceeding further. In particular, you should be able to start the program, and open and save *PSY* input files.

Example 2.1 Planned orthogonal contrasts

Suppose that the researchers running the Depression experiment decide to construct a 95% individual CI on each of $(J-1)=3$ planned $\{m, r\}$ contrasts with the following coefficient vectors:

$$\mathbf{c}_1' = \left[0.\dot{3}\ \ 0.\dot{3}\ \ 0.\dot{3}\ \ -1\right] \qquad \left(\psi_1 = \frac{\mu_1+\mu_2+\mu_3}{3}-\mu_4\right)$$

$$\mathbf{c}_2' = \left[\,0.5\ \ 0.5\ \ -1\ \ 0\,\right] \qquad \left(\psi_2 = \frac{\mu_1+\mu_2}{2}-\mu_3\right)$$

and

$$\mathbf{c}_3' = \left[\,1\ \ -1\ \ 0\ \ 0\,\right] \qquad \left(\psi_3 = \mu_1-\mu_2\right).$$

These coefficient vectors are uncorrelated with each other, and they therefore define *orthogonal* contrasts if sample sizes (n_j) are equal.[8] Orthogonal contrasts are linearly independent, and their sample values are statistically independent. As we will see, orthogonal contrasts partition between-cells variation into additive components.

A *PSY* analysis requires an input file containing contrast coefficient vectors and data. The contrasts section of the input file follows.

```
[BetweenContrasts]
1  1  1 -3 Ts - C
1  1 -2  0 NT,ST - MCT
1 -1  0  0 NT - ST
```

All contrast coefficients must be integers. It is easier to enter the coefficients of $3\mathbf{c}_1' = [1\ 1\ 1\ {-3}]$ than it is to enter $[0.33333\ 0.33333\ 0.33333\ {-1}]$, the mean difference coefficient vector. *PSY* transforms all contrasts into mean difference contrasts unless requested not to do so on the *Analysis Options* menu (see Appendix A). Each coefficient vector is entered as a row vector, with coefficients separated by at least one space. An optional 12 column (or shorter) label for each contrast may be entered following the final coefficient. A space must separate the label from the final coefficient.

Data are entered under the heading [Data] with one row per subject, beginning with a subject in group 1, followed by the remaining subjects in that group. For every subject the first entry is the value of a group membership variable specifying the number of the group (1, 2, 3, ...) to which that subject belongs. The second (and in this case last) value in each is the dependent variable (Y) score, separated from the group-membership score by a space. The data section of the file *depression.in* contains $N = 80$ rows of data , as follows:

```
[Data]
1 30
  .
  .
1 27
2 29
  .
  .
2 32
3 13
  .
  .
3 12
4 22
  .
  .
4  4
```

The default settings for a *PSY* analysis produce output including the following:

```
Means and Standard Deviations
 Group 1
  Mean          26.850
  SD             9.063
 Group 2
  Mean          26.650
  SD             8.616
 Group 3
  Mean          23.950
  SD             8.287
 Group 4
  Mean          17.100
  SD             7.454
             Analysis of Variance Summary Table

             Source          SS       df         MS            F
             ------------------------------------------------
             Between
             ------------------------------------------------
Ts - C       B1         1139.704      1    1139.704       16.245
NT,ST - MCT  B2          104.533      1     104.533        1.490
NT - ST      B3            0.400      1       0.400        0.006
             Error      5331.850     76      70.156
             ------------------------------------------------
```

```
            Individual 95% Confidence Intervals
            -----------------------------------
            The CIs refer to mean difference contrasts,
             with coefficients rescaled if necessary.
             The rescaled contrast coefficients are:

            Rescaled Between contrast coefficients
       Contrast         Group...
                          1           2           3           4
Ts - C          B1      0.333       0.333       0.333      -1.000
NT,ST - MCT     B2      0.500       0.500      -1.000       0.000
NT - ST         B3      1.000      -1.000       0.000       0.000
            Raw CIs (scaled in Dependent Variable units)
            ------------------------------------------------------
            Contrast   Value        SE           ..CI limits..
                                               Lower       Upper
            ------------------------------------------------------
Ts - C          B1       8.717       2.163       4.409      13.024
NT,ST - MCT     B2       2.800       2.294      -1.769       7.369
NT - ST         B3       0.200       2.649      -5.075       5.475
            ------------------------------------------------------

          Approximate Standardized CIs (scaled in Sample SD units)
            ------------------------------------------------------
            Contrast   Value        SE           ..CI limits..
                                               Lower       Upper
            ------------------------------------------------------
Ts - C          B1       1.041       0.258       0.526       1.555
NT,ST - MCT     B2       0.334       0.274      -0.211       0.880
NT - ST         B3       0.024       0.316      -0.606       0.654
            ------------------------------------------------------
```

Suppose that a mean difference of more than 0.5 standard deviation units is regarded as clinically important (or clinically significant, as distinct from statistically significant), a difference of less than 0.2 standard deviation units is deemed to be trivially small, and the importance of a difference between these two values is difficult to assess. Given these guidelines, the standardized CIs imply that treatment for depression (ignoring differences between treatments) has a clinically important positive effect (ψ_1), but it is not clear whether there are important differences between treatments in the magnitudes of their effects. It is clear only that extended treatment (ignoring the nature of the treatment) is either superior to or practically equivalent to minimal-contact treatment (ψ_2). Although the midpoint of the CI on the difference between the effects of the new and standard treatments is very close to zero ($\hat{\psi}_3/\hat{\sigma}_\varepsilon = 0.024$), the CI limits show that this estimate is not sufficiently precise to exclude the possibility of a nontrivial difference in either direction [$\psi_3/\sigma_\varepsilon \in (-0.606, 0.654)$].

Comments on the analysis This analysis is based on a set of *Helmert* contrasts in which the average of the first $(J - 1)$ means is compared with the last, the average of the first $(J - 2)$ is compared with the second last, and so on. Helmert contrasts are *orthogonal* in equal-n experiments. Any set of $(J - 1)$ orthogonal contrasts partitions the sum of squares between groups into additive components, where the components are the *contrast sums of squares* shown in the *PSY* ANOVA summary table.[9] In this case

$$\begin{aligned} SS_B &= SS(\hat{\psi}_1) + SS(\hat{\psi}_2) + SS(\hat{\psi}_3) \\ &= 1139.704 + 104.533 + 0.400 \\ &= 1244.637. \end{aligned}$$

The contrast *F* statistics in the final column of the ANOVA summary table (not to be confused with the overall ANOVA *F* statistic) are usually reported in planned contrasts analyses based on significance tests, but they are not used in central CI analyses. The *F* statistic for a planned contrast is distributed as a noncentral *F* distribution with $\nu_1 = 1$, whereas the ANOVA *F* statistic (not reported in the *PSY* summary table) is distributed as a noncentral *F* distribution with $\nu_1 = (J - 1)$. Any *F* statistic with $\nu_1 = 1$ is the square of a *t* statistic; the sign of a contrast *t* statistic is the same as the sign of the sample value of the contrast. We can tell from the *PSY* output that the *t* statistic for ψ_1 is $+\sqrt{16.245} = 4.031$. The exact noncentral standardized CI on ψ_1 (obtained from the method outlined in Appendix C) is $\psi_1/\sigma_\varepsilon \in (0.505, 1.570)$. This result illustrates the fact that approximate standardized CIs are too narrow (only slightly too narrow in this case) when estimated effect sizes are large. Exact standardized CIs on the remaining contrasts are $\psi_2/\sigma_\varepsilon \in (-0.206, 0.873)$ and $\psi_3/\sigma_\varepsilon \in (-0.596, 0.644)$.

Example 2.2 Bonferroni-t confidence intervals on contrasts

Selection of *Bonferroni t* from the Analysis Options menu (discussed in Appendix A) produces a set of simultaneous CIs controlling the FWER, rather than individual CIs controlling the PCER, as shown in the following excerpt from the output.

```
                Bonferroni 95% Simultaneous Confidence Intervals
                ------------------------------------------------

                Raw CIs (scaled in Dependent Variable units)
                ----------------------------------------------------
                Contrast   Value        SE          ..CI limits..
                                                   Lower       Upper
                ----------------------------------------------------
Ts - C          B1         8.717       2.163       3.422      14.011
NT,ST - MCT     B2         2.800       2.294      -2.816       8.416
NT - ST         B3         0.200       2.649      -6.284       6.684
                ----------------------------------------------------

            Approximate Standardized CIs (scaled in Sample SD units)
                ----------------------------------------------------
                Contrast   Value        SE          ..CI limits..
                                                   Lower       Upper
                ----------------------------------------------------
Ts - C          B1         1.041       0.258       0.409       1.673
NT,ST - MCT     B2         0.334       0.274      -0.336       1.005
NT - ST         B3         0.024       0.316      -0.750       0.798
                ----------------------------------------------------
```

Example 2.3 Scheffé post hoc analysis

Selection of the *post hoc* option from the Analysis Options screen produces a *PSY* analysis with Scheffé SCIs on the contrasts in the input file. In the following analysis of the depression data, the contrasts are typical of those that might have been chosen after a preliminary examination of the sample means. The contrasts section of the input file follows.

```
[BetweenContrasts]
1  1  0 -2 NT,ST - C
1 -1  0  0 NT - ST
1  1 -2  0 NT,ST - MCT
0  0  1 -1 MCT - C
```

Excerpts from the output file follow.

```
                    Post hoc 95% Simultaneous Confidence Intervals
                    ----------------------------------------------
                    The CIs refer to mean difference contrasts,
                     with coefficients rescaled if necessary.
                     The rescaled contrast coefficients are:

                    Rescaled Between contrast coefficients
          Contrast            Group...
                                1            2            3            4
NT,ST - C       B1            0.500        0.500        0.000       -1.000
NT - ST         B2            1.000       -1.000        0.000        0.000
NT,ST - MCT     B3            0.500        0.500       -1.000        0.000
MCT - C         B4            0.000        0.000        1.000       -1.000

                    Raw CIs (scaled in Dependent Variable units)
                    ---------------------------------------------------
                    Contrast     Value          SE          ..CI limits..
                                                          Lower        Upper
                    ---------------------------------------------------
NT,ST - C       B1               9.650        2.294        3.092       16.208
NT - ST         B2               0.200        2.649       -7.373        7.773
NT,ST - MCT     B3               2.800        2.294       -3.758        9.358
MCT - C         B4               6.850        2.649       -0.723       14.423
                    ---------------------------------------------------

                 Approximate Standardized CIs (scaled in Sample SD units)
                    ---------------------------------------------------
                    Contrast     Value          SE          ..CI limits..
                                                          Lower        Upper
                    ---------------------------------------------------
NT,ST - C       B1               1.152        0.274        0.369        1.935
NT - ST         B2               0.024        0.316       -0.880        0.928
NT,ST - MCT     B3               0.334        0.274       -0.449        1.117
MCT - C         B4               0.818        0.316       -0.086        1.722
                    ---------------------------------------------------
```

The experimenter could conclude from ψ_1 that extended treatment (ignoring the nature of the treatment) has a positive effect (perhaps a large effect), and that the effect size has been estimated with poor precision. The analysis permits no informative inference to be made about the direction or magnitude of the difference between the new and standard treatments (ψ_2). Similarly, the analysis has very little to say about the difference between extended treatment and minimal-contact treatment (ψ_3). Minimal-contact treatment may well be superior (perhaps substantially so) to no treatment, but the possibility that it is practically equivalent to no treatment cannot be excluded (ψ_4).

Comments on the analysis The Scheffé CC for this analysis is

$$\sqrt{\nu_B F_{\alpha;\nu_B,\nu_\varepsilon}} = \sqrt{3\times F_{.05;3,76}} = \sqrt{3\times 2.725} = 2.859$$

which is 16.8% larger than the Bonferroni-*t* CC used in Example 2.2, and 43.5% larger than for the planned orthogonal contrasts analysis (Example 2.1). It is therefore not surprising that this analysis appears to say more about the lack of precision of estimates of effect sizes than it does about their magnitudes. The flexible basis provided by the Scheffé procedure for the choice of contrasts can sometimes offset the disadvantages associated with relatively poor precision. For an example, see Question 1 at the end of this chapter.

Note that because the Scheffé CI for ψ_1 implies that population means (and effect parameters) are heterogeneous, it also implies that the lower limit of the 90% CI on f must be greater than zero, and that the .05-level ANOVA F test must reject the homogeneity hypothesis. Given the outcomes of the Scheffé analysis, the F test is redundant.

Despite problems with precision, the Scheffé analysis provides much more information than an estimate of an overall index of effect size such as the 90% CI on Cohen's f [$f \in (0.237, 0.641)$] reported earlier. In principle, an interval estimate of f can be useful if it suggests that the ANOVA model can be replaced with a no-effects model ($Y_{ij} = \mu + \varepsilon_{ij}$), thereby removing the need (and the basis) for a contrasts analysis. In practice, the possibility of such an outcome can usually be ignored, unless the overall sample size is very large.

Alternatives to Bonferroni-*t* SCIs for restricted analyses

The Bonferroni-*t* procedure always controls the FWER in analyses restricted to k planned contrasts, but it does so conservatively. In some restricted analyses (usually analyses restricted to comparisons when sample sizes are equal), it is possible to choose an optimal CC that controls the FWER exactly, thereby providing SCIs with greater precision than Bonferroni-*t* SCIs.

When sample sizes are equal, the *Tukey* procedure (Tukey, 1953/1994) controls the FWER exactly in analyses based on a complete set of $J(J-1)/2$ comparisons. Raw (but not standardized) Tukey SCIs can be obtained from *SPSS* and a number of other statistical packages. *PSY* does not support the Tukey procedure. If you need to use *PSY* to construct Tukey SCIs (perhaps because you want standardized CIs), select *User-supplied CCs* from the Analysis Options screen, and insert the relevant value of $|q^*|_{\alpha;J,J(n-1)}$ which can be obtained from Table F1 in Appendix F.[10] If you want to control the

FWER at $\alpha = .05$, the required CC is $|q^*|_{.05;4,76} = 2.627$. (You can obtain this CC by interpolating between the values given in Table F1(a) for $|q^*|_{.05;4,70}$ and $|q^*|_{.05;4,80}$.) An excerpt from the *PSY* output for this analysis is shown below.

Special Confidence Intervals: User-supplied Critical Constants
Between main effect CC: 2.627

Raw CIs (scaled in Dependent Variable units)

	Contrast	Value	SE	..CI limits.. Lower	Upper
NT-ST (1-2)	B1	0.200	2.649	-6.758	7.158
NT-MCT (1-3)	B2	2.900	2.649	-4.058	9.858
NT-C (1-4)	B3	9.750	2.649	2.792	16.708
ST-MCT (2-3)	B4	2.700	2.649	-4.258	9.658
ST-C (2-4)	B5	9.550	2.649	2.592	16.508
MCT-C (3-4)	B6	6.850	2.649	-0.108	13.808

Approximate Standardized CIs (scaled in Sample SD units)

	Contrast	Value	SE	..CI limits.. Lower	Upper
NT-ST (1-2)	B1	0.024	0.316	-0.807	0.855
NT-MCT (1-3)	B2	0.346	0.316	-0.484	1.177
NT-C (1-4)	B3	1.164	0.316	0.333	1.995
ST-MCT (2-3)	B4	0.322	0.316	-0.508	1.153
ST-C (2-4)	B5	1.140	0.316	0.309	1.971
MCT-C (3-4)	B6	0.818	0.316	-0.013	1.649

Because raw Tukey SCIs control the FWER exactly, any other SCI procedure necessarily produces more conservative raw SCIs for this particular analysis. Bonferroni-*t* SCIs (with a CC of $t_{.05/(2\times6);76} = 2.709$) are 3.1% wider than the Tukey SCIs shown above, so the increase in precision provided by the Tukey procedure is modest rather than dramatic in this particular case.

The procedure developed by Dunnett (1955) is another example of an SCI procedure that is optimal for a particular type of planned analysis. Dunnett's procedure controls the FWER exactly when each of $(J-1)$ treatment means is compared with a single control mean when sample sizes are equal. In an analysis of the Depression data restricted to the three comparisons comparing each of the three treatment means with the final control mean, Bonferroni-*t* SCIs are 2.1% wider than Dunnett SCIs.

These examples illustrate the fact that, although the Bonferroni-*t* procedure is always conservative (in the sense that it produces raw SCIs that are always slightly wider than is necessary to control the FWER at exactly α), it is often only slightly conservative.

The Bonferroni-*t* procedure has at least one competitor that always produces a smaller critical value in any planned analysis. The Sidak-*t* procedure (Sidak, 1967) controls the FWER by setting the PCER at $1-(1-\alpha)^{1/k}$, which is always slightly larger than the Bonferroni-*t* PCER of α/k. The Sidak-*t* CC is therefore always smaller than the corresponding Bonferroni-*t* CC. In practice, however, the SCIs produced by these procedures are virtually indistinguishable. The Bonferroni-*t* intervals in Example 2.2, for example, are only 0.3% wider than Sidak-*t* intervals. The Bonferroni approach to error-rate control in planned analyses is well known, has historical precedence over Sidak's approach, and in any event is almost as efficient as Sidak's approach.[11] For these reasons, the Sidak-*t* procedure is not discussed further.

Further reading

Many books provide a much more detailed account of the traditional null hypothesis significance testing approach to one-way ANOVA than that given here. Of these, the book by Harris (1994) is highly recommended if you are looking for a treatment of ANOVA based on a unified approach that is compatible with the treatment in this book, except for an emphasis on directional inference rather than CI inference. Harris does discuss the role of CIs in cases where directional inference is not possible.

If you are interested in working through the derivations of ANOVA procedures, you will probably enjoy the treatments by Hays (1981) or Kirk (1995). Timm (1975) provides an excellent treatment for those who feel comfortable with matrix algebra.

Most of the standard references on ANOVA are large books dealing primarily with experimental design, such as Kirk (1995), Maxwell and Delaney (1990) and Winer, Brown and Michels (1991). These books contain a wealth of information on topics related to ANOVA, but only brief discussions of CI inference.

Questions and exercises

1. An experiment is designed to assess the effect of requiring payment for a treatment for arachnophobia (fear of spiders). One hundred arachnophobic participants are randomly assigned to one of five conditions ($n = 20$):

Condition 1: Free treatment
Condition 2: \$100 fee, to be refunded if treatment is unsuccessful
Condition 3: \$200 fee, to be refunded if treatment is unsuccessful
Condition 4: \$100 fee, regardless of treatment outcome
Condition 5: \$200 fee, regardless of treatment outcome

All participants are made aware of the fee (if any) and contingency (if any) associated with their own treatment condition (but not other conditions) before treatment commences. Outcome is assessed on an arachnophobia scale where a relatively high score would indicate a relatively high phobia level. A difference of four points or more on this scale is regarded as a clinically significant difference.

(a) Plan a set of contrasts addressing the following questions:

- (i) What is the average effect of charging a fee?
- (ii) What is the average effect of increasing the fee from $100 to $200?
- (iii) What is the average effect of making a fee contingent on outcome?
- (iv) Does the size of the contingency effect vary across fee levels?

(b) Given the data in the file *Ch2 Q1b.in*, use *PSY* to construct interval estimates of these contrasts, controlling the PCER at $\alpha = .05$. Interpret the outcome of the analysis.

Hint: Only three of the four contrasts should be scaled as mean difference contrasts. Since *PSY* uses the same scaling for all contrasts in one analysis, you will need to run two analyses, selecting a different scaling option for the second run.

(c) Suppose now that the researcher was interested only in confident direction inferences. What directional inferences would follow from the CIs you have constructed (or from significance tests compatible with these CIs)?

(d) Carry out another CI analysis of the data, this time controlling the FWER rather than the PCER. (Treat the analysis as planned.) Comment on the consequences of changing the error-rate unit.

(e) Now carry out a similar analysis (the same planned contrasts, FWER) of the different data set in the file *Ch2 Q1e.in*. Does this analysis appear, after the event, to be more or less satisfactory than the previous analysis? Why?

(f) Carry out a post hoc analysis of the data in *Ch2 Q1e.in*. Feel free to base the analysis on any contrasts that appear to be sensible, given the experimental design *and* the data. You may wish to include some of the contrasts used to answer earlier parts of this question. Does this analysis appear to be more or less satisfactory than the previous planned analysis? Why?

(g) Can you think of any problems associated with a post hoc analysis carried out only because a planned analysis was deemed to be unsatisfactory, given the way the data turned out?

Notes

1. These are the standard assumptions underlying traditional ANOVA-model analyses. Alternative *robust* analyses are designed to avoid the need to assume that error is

normally distributed with homogeneous variances. In general, robust analyses are computer intensive and not currently readily accessible to most researchers.

2. This is an unusual usage of the term 'standard deviation', which usually refers to variability in the values of a random variable, rather than to variability in a set of fixed constants such as effect parameters.

3. The f parameter cannot be negative, and the standard F test is therefore a one-tailed test of the homogeneity hypothesis ($f = 0$) against a directional alternative ($f > 0$), using only the upper tail of the relevant central F distribution. The *lower* limit of the 90% two-sided CI [equivalent to the single limit of the 95% single-sided CI (ll, ∞)] is greater than zero if and only if $p < .05$ [where p is the p value (given $f = 0$) associated with the F statistic].

4. A mean difference contrast is sometimes defined as a contrast with coefficients whose absolute values sum to 2 ($\sum|c_j| = 2.0$). This definition is equivalent to that given here ($\sum c_j^+ = 1.0$) when applied to contrast coefficient vectors, but not to some other coefficient vectors we will be concerned with in Chapter 7.

5. When sample sizes are equal, the effective sample size n^* available for the estimation of an $\{m, r\}$ contrast is the harmonic mean of mn and rn. For a comparison, $n^* = n$. For a complex $\{m, r\}$ contrast, $n^* > n$.

6. When sample sizes are equal, the coefficients of the maximal contrast are proportional to deviation means $(M_j - M)$.

7. *SPSS* can also produce CI limits on δ, given the syntax available from Michael Smithson's web page. See Appendix C for details.

8. In an equal-n experiment two contrasts are mutually orthogonal if the sum of products of their coefficients (and therefore their correlation) is zero.

9. The sum of squares for a contrast is $SS(\hat{\psi}_g) = \hat{\psi}_g^2 \big/ \Sigma_j\left(c_{gj}^2/n_j\right)$.

10. Following Hsu (1996), CCs for the Tukey procedure are expressed here (and in Table F1 in Appendix F) as critical values of $|q^*| = q/\sqrt{2}$, where q is the studentized range distribution.

11. In a planned analysis the Bonferroni-t procedure controls the *per-family error rate* (PFER), defined as the expected number of noncoverage errors in a family of k CIs, exactly (PFER = α). Unlike the FWER, the PFER treats a set of inferences containing multiple errors as a worse outcome than a set of inferences containing only one error, and for this reason is sometimes preferred as a family-based error-rate criterion (Klockars and Hancock, 1994). Those who prefer to control the PFER would regard the Bonferroni-t procedure as the optimal procedure for multiple comparisons. From this perspective, the Sidak-t, Dunnett and Tukey procedures are (slightly) excessively liberal (PFER > α). We will find it useful to control the PFER in some non-standard analyses discussed in Chapters 4, 5, 6 and 7.

3 Precision and Power

One of the important features of a CI analysis is the information provided about precision of estimation. If contrast values are estimated with a high degree of precision, CIs are narrow and inferences are unequivocal. Very high precision leads to the possibility of practical equivalence inference, an important type of confident inference that is not possible when a CI is wider than the relevant equivalence interval. Even if very high precision is not realistically attainable, any well-designed experiment should be capable of producing sufficiently precise estimates of contrast values to ensure that CIs will be genuinely informative. Experiments are usually costly to run, and there is little point in running an experiment with almost no chance of providing informative answers to the questions it is designed to address.

The Scheffé analysis of the Depression data (Example 2.3) illustrates the problems arising from low precision. The width of a 95% Scheffé CI on a comparison such as ψ_2 is $1.808\,\hat{\sigma}_\varepsilon$, leaving open the possibility of substantial differences in the interpretation of different values in the interval. In the case of ψ_2, for which the point estimate of the standardized population value is very close to zero ($\hat{\psi}_2/\hat{\sigma}_\varepsilon = 0.024$), the standardized CI (–0.880, 0.928) includes large differences in both directions, and is therefore almost completely uninformative. (The interval does exclude extremely large differences, and is therefore not absolutely uninformative.)

In principle, precision of estimation can be controlled by selecting an appropriate sample size. In practice, the experimenter is unlikely to have access to the resources required to implement a design requiring (say) several thousand subjects, but it is still useful to be able to determine in advance the precision of estimation based on a sample of any given size. As we will see, it is easier to determine precision than it is to determine statistical power, the analogous concept required for sample size determination for confident directions inference and inference at lower levels.

Factors influencing precision

The precision of an interval estimate of the value of a contrast is inversely related to the CI half-width (w). We have already seen that w decreases (and therefore precision increases) as noncoverage error rate, sample size and contrast complexity increase. The half-width of raw CIs (but not standardized CIs) also decreases as error variance (σ_ε^2) decreases, via the effect of error variance on the standard error of contrasts. The final factor influencing w is the procedure used to construct the interval (t, Bonferroni-t or Scheffé). In the case of the Bonferroni-t procedure, w increases as the number of planned contrasts (k) increases.

The effect of error rate (and therefore confidence level) on precision can be seen by reanalysing the Depression data with a much higher error rate than that used in the analyses reported earlier. This analysis uses a CC of 1.0, so that the half-width of each CI is the estimated standard error of the relevant contrast. The reason for choosing such a small CC is that in some areas of research it is a common practice to base informal inferences on figures showing sample means as midpoints of intervals with half-widths of one standard error. A CC of 1.0 produces a PCER of .32 in a planned analysis of the Depression data, and it allows the FWER to reach .80 in a post hoc analysis. Part of the *PSY* output from this analysis (obtained by providing a user-supplied CC of 1.0) is shown below.

Approximate Standardized CIs (scaled in Sample SD units)

	Contrast	Value	SE	..CI limits.. Lower	Upper
NT,ST-C	B1	1.152	0.274	0.878	1.426
NT-ST	B2	0.024	0.316	-0.292	0.340
NT,ST-MCT	B3	0.334	0.274	0.060	0.608
MCT-C	B4	0.818	0.316	0.502	1.134

If the smallest clinically significant difference is $0.4\,\sigma_\varepsilon$ then the inferences from these intervals are for the most part quite clear-cut. The average of the treatment means is substantially greater than the control mean (ψ_1), and the difference in the effectiveness of the two treatments is not clinically significant (ψ_2). The average of the treatment means is also greater than the minimal-contact mean, but perhaps trivially so (ψ_3). Minimal contact has a clinically significant effect, perhaps a large effect (ψ_4). These inferences are clear not because the estimated contrast values are large (although some are), but because the intervals are narrow, relative to the 95% simultaneous intervals discussed earlier. The cost of this apparent increase in precision is the increase in the nominated FWER from a conventionally conservative level of $\alpha = .05$ to an extremely liberal level of .80. Although this level of FWER 'control' has not been recommended by anyone, some researchers would be satisfied with a procedure that appears to produce a PCER of .32. (It must be remembered, of course, that the contrasts in this analysis are not linearly independent, nor are they defined independently of the data.)

In principle, problems with precision can manifest themselves in one of two ways: through a combination of low error rates (high confidence levels) and wide CIs, or through a combination of high error rates (low confidence levels) and narrow CIs. The first approach is (or should be) preferred by those who believe that conventional α levels are generally appropriate for null hypothesis significance tests. It is obviously desirable to be able to produce inferences with low error rates, but it is not clear that low error rates for relatively uninformative inferences are always preferable to higher error rates for more informative inferences.

Ideally, of course, it should be possible to maintain low error rates and at the same time produce informative inferences. The most obvious way of achieving this ideal is to increase the sample size, thereby decreasing the standard error of contrast estimates. How, then, can an experimenter choose a sample size to ensure that the resulting CIs are sufficiently narrow that they can be interpreted more or less unequivocally while maintaining an FWER of .05?

Precision of estimation with known error variance

It will be convenient initially to simplify the discussion by supposing that the error variance is known. With known error variance, CIs can be constructed from

$$\psi_g \in \hat{\psi}_g \pm \text{CC} \times \sigma_{\hat{\psi}_g}$$

which makes use of the (known) standard error of each contrast, rather than the estimated standard error used by (2.16). The CC for any particular CI procedure is independent of sample size. As a result, CI half-widths ($w_g = \text{CC} \times \sigma_{\hat{\psi}_g}$) and the sample size required to achieve these half-widths can be determined before the experiment is run. The sample size required to achieve a given value of w_g is

$$n = \sum_j c_{gj}^2 \left(\frac{\text{CC} \times \sigma_\varepsilon}{w_g} \right)^2 . \tag{3.1}$$

If the intention is to ensure that no mean difference contrast has a half-width in excess of a specified value, then n should be determined for comparisons, for which $\sum c^2 = 2$. If the half-width of intervals on comparisons is expressed in standard deviation units, (3.1) simplifies to

$$n = 2 \left(\frac{\text{CC}}{w_{z\{1,1\}}} \right)^2 , \tag{3.2}$$

where $w_{z\{1,1\}}$ is the required comparison half-width.

Suppose that the Depression experiment was to be repeated, with the sample size chosen to produce 95% post hoc CIs on mean difference contrasts with a maximum half-width of 0.2 standard deviation units. This specification is chosen to ensure that a CI including the value zero cannot also include clinically significant differences. If σ_ε^2 is known to the experimenter, the required CC is

$$\text{CC} = \sqrt{3F_{.05;3,\infty}} = \sqrt{\chi^2_{.05;3}} = 2.795.$$

(Note that $F_{\nu_1,\infty} = \chi^2_{\nu_1}/\nu_1$.) This is a slightly smaller CC than that used in the post hoc analysis of the Depression data set discussed in Chapter 2 ($\sqrt{3F_{.05;3,76}}$ = 2.859), because in that analysis it was not assumed that σ_ε^2 was known to the experimenter. Given $w_{z_{\{1,1\}}} = 0.2$ and a CC of 2.795, we can use (3.2) to find the appropriate sample size:

$$n = 2\left(\frac{2.795}{0.2}\right)^2 = 390.6.$$

The experimenter would need a total of $N = Jn = 4 \times 391 = 1564$ subjects in the experiment to ensure that the widest 95% post hoc CI on a mean difference contrast would have a half-width no greater than 0.2 standard deviation units. It is not surprising, then, that CIs based on a total of $N = 80$ subjects produced an unsatisfactory set of inferences.

Before we abandon the unrealistic assumption that the experimenter knows the population variance, we will examine some of the other factors that influence CI width. The effect of contrast complexity can be seen by comparing the half-width of the interval on a {2, 2} contrast like ψ_4 (where $\mathbf{c}_4' = [0.5\ 0.5\ -0.5\ -0.5]$) with that on a comparison. Because the value of $\sum c^2$ (1.0) is half that for a comparison, the sample size required to achieve a required level of precision from an interval estimate of a {2, 2} contrast is exactly half that required to achieve the same precision from an interval estimate of a comparison. If all of the contrasts to be examined were complex contrasts, substantial savings in required sample size could be achieved. Given the design we are discussing, however, increasing the overall level of contrast complexity can be achieved only by excluding some contrasts of interest from the analysis.

Choosing an analysis strategy

The choice of analysis strategy (deciding whether to restrict the analysis to a set of planned contrasts) can have a substantial effect on the half-width of CIs. If a decision is made before running a four-group experiment to analyse the data with a set of three planned contrasts, the minimum number required to account for all between-group variation, then the CC for 95% Bonferroni-*z* intervals (appropriate if the error variance is known) will be $z_{.05/(2\times3)} = 2.394$. If at least one of the planned contrasts is a comparison, then the sample size required to achieve a maximum half-width of $0.2\,\sigma_\varepsilon$ is $n = 2(2.394/0.2)^2 = 286.6$. Rounding up (to 287) gives a required N of $4 \times 287 = 1148$, 73.4% of the sample size required to achieve the same maximum width from post hoc intervals. The cost of this reduction in required sample size is a loss of flexibility in the analysis. Before deciding to carry out a restricted analysis, the experimenters should satisfy themselves that the contrasts planned for the experiment can produce a satisfactory analysis even if the pattern of means does not turn out as expected.

The choice of error-rate unit (the individual contrast or the set of contrasts) can have a very large effect on the required sample size. If 95% individual CIs were to be constructed, then the relevant CC (assuming known error variance) would be $z_{.05/2} = 1.960$, so the required sample size would be $n = 2(1.96/0.2)^2 = 192.1$. Rounding up gives a required N of $4 \times 193 = 772$, 67.2% of the sample size required for Bonferroni-z intervals, and only half the sample size required for post hoc intervals.

If an experimenter wishes to know in advance the half-width of CIs for a given sample size, the relevant equation for raw intervals is

$$w_g = \text{CC} \times \sigma_\varepsilon \sqrt{\frac{\sum_j c_{gj}^2}{n}}. \tag{3.3}$$

The half-width of standardized intervals is obtained by deleting σ_ε from (3.3).

Precision of estimation with unknown error variance

When error variance is unknown, as is almost always the case in practice, the half-width of raw CIs cannot be calculated before the experiment is run. If a reasonable estimate of the population standard deviation can be obtained from other sources, then the estimated value can be substituted for σ_ε in (3.3) to produce a reasonable estimate of w.

The half-width of approximate standardized CIs can be calculated before an experiment with a given sample size is run, from

$$w_{\hat{z}} = \text{CC} \sqrt{\frac{\sum_j c_{gj}^2}{n}}. \tag{3.4}$$

Thus it is possible to determine in advance that with $J = 4$ and $n = 20$, the half-width of an approximate standardized 95% Scheffé interval (with CC = 2.859) for a comparison will be $w_{\hat{z}_{\{1,1\}}} = 2.859 \times \sqrt{2/20} = 0.904$, exactly the half-width of the standardized Scheffé intervals on comparisons in Example 2.3. Given the half-width of intervals on comparisons, the half-width of the interval on any complex contrast ψ_h can be calculated from

$$w_{\hat{z}_h} = w_{\hat{z}_{\{1,1\}}} \sqrt{\frac{\sum_j c_{hj}^2}{2}} . \tag{3.5}$$

Thus if $w_{\hat{z}_{\{1,1\}}} = 0.904$, the half-width of the {2, 1} intervals on ψ_1 and ψ_3 (with $\sum c^2 = 1.5$) is

$$w_{\hat{z}_{\{2,1\}}} = 0.904\sqrt{\frac{1.5}{2}} = 0.783.$$

More generally, if the half-width w_g of the CI on any contrast ψ_g is known, the half-width of the interval on any other contrast ψ_h can be calculated from

$$w_h = w_g \sqrt{\frac{\sum_j c_{hj}^2}{\sum_j c_{gj}^2}} \tag{3.6}$$

How many subjects? Choosing a sample size to produce a desired value of $w_{\hat{z}}$ when the error variance is unknown is theoretically more complicated than choosing a sample size to produce a desired value of w_z when the error variance is known. If we wish to use an expression like (3.1) to solve for n before the experiment is run, we cannot calculate the value of CC, which depends on the number of degrees of freedom for error and therefore on the unknown value of n. It turns out, however, that for conventional α levels a good estimate of the required n can be obtained from

$$n = 1 + \sum_j c_{gj}^2 \left(\frac{\text{CC}}{w_{\hat{z}_g}}\right)^2 , \tag{3.7}$$

where CC is calculated on the assumption that the error variance is known, and n is always rounded up to the next highest integer.

Suppose, for example, that we wish to know the n required to produce an approximate standardized Bonferroni-t 95% CI with a half-width of 0.50 for a comparison when $J = 4$ and $k = 3$. The CC required for (3.7) is $t_{.05/(2\times3);\infty} = z_{.05/(2\times3)} = 2.394$, so $n = 1 + 2(2.394/0.5)^2 = 46.8$. Rounding up, we have $n = 47$ and $N = 188$.

If $n = 47$, then $\nu_E = 184$ and the CC for Bonferroni-t 95% CIs is $t_{.05/(2\times3);184} = 2.416$, so that [according to (3.4)] the half-width of intervals on comparisons must be $w_{\hat{z}_{\{1,1\}}} = 2.416 \times \sqrt{2/47} = 0.50$, the required value.

Sample size tables The same result can be obtained without calculations from Table F3 in Appendix F. Enter that table with $k = 3$ and $w = 0.5$ to read off the tabled value of $n = 47$. Tables F2 to F4 [constructed from (3.7)] show n as a function of $w_{\hat{z}}$ for approximate Tukey, Bonferroni-*t* and Scheffé standardized CIs on comparisons for $\alpha = .05$ and .10. Values for individual CIs can be obtained from the first column (headed $J = 2$) of Table F2. Values for complex contrasts in Bonferroni-*t* and Scheffé analyses are obtained by dividing the tabled value by $2/\sum c^2$.

Example 3.1 Controlling precision

An experiment with $J = 5$ groups is designed to evaluate $k = 4$ planned $\{m, r\}$ contrasts: one $\{4, 1\}$ contrast, one $\{2, 2\}$ contrast and two comparisons. The experimenters would like to be able to construct 90% SCIs with a maximum half-width of 0.3 standard deviation units.

The Bonferroni-*t* procedure will produce narrower CIs than the Scheffé procedure, because k is not greater than ν_B. Table F3 shows that when $k = 4$ and $w_{\hat{z}} = 0.3$, the sample size required for Bonferroni-*t* intervals on comparisons is $n = 113$ ($N = Jn = 565$ subjects in all).

If the experimenters were unwilling or unable to specify an N of this magnitude, they could consider the consequences of using smaller sample sizes. The sample size required to produce $w_{\hat{z}} = 0.3$ for the CI on the contrast with the largest effective sample size (the $\{2, 2\}$ contrast with four coefficients of ± 0.5 and $\sum c^2 = 1.0$) is obtained by dividing the tabled n of 113 by $2/\sum c^2 = 2.0$. The resulting overall sample size is $N = 285$ ($n = 57$, after rounding up from $113/2 = 56.5$). Half-widths of the remaining CIs can be estimated from (3.6). For the $\{4, 1\}$ contrast (with four coefficients of ± 0.25 and one coefficient of ± 1.0), $\sum c^2 = 1.25$. It follows from (3.6) that if the half-width of a CI on a contrast with $\sum c^2 = 1.0$ is 0.3, then the half-width of an interval on a contrast with $\sum c^2 = 1.25$ must be $0.3 \times \sqrt{1.25/1} = 0.335$. Similarly, the half-width of an interval on a comparison must be $0.3 \times \sqrt{2/1} = 0.424$.

Power

The power $(1-\beta)$ of a significance test is the probability of correctly rejecting a null hypothesis when it is false. The complement of power (β) is the probability of (incorrectly) failing to reject a false null hypothesis. Power analysis is directly relevant to directional and inequality inference rather than CI inference, because CI inference is not directly concerned with significance testing. Nevertheless, CI inference can imply directional inference, so power analysis (as well as precision analysis) can be of interest to experimenters who intend to carry out an ANOVA via CIs rather than significance tests.

Suppose that the experimenters planning the Depression experiment want to select a sufficiently large sample size to ensure that if the effect of the new treatment differs from the effect of the standard treatment by $0.5\sigma_\varepsilon$, then the probability of a directional inference from a 95% individual CI on this comparison will be .8. In the language of significance testing, the experimenters want the power of the test of $H_0\colon \psi_3 = 0$ to be $(1-\beta) = .8$ if the alternative hypothesis $H_1\colon |\psi_3| = 0.5\sigma_\varepsilon$ is true. We will refer to this type of power as *conditional* power, because it is conditional on the effect size specified by a particular alternative hypothesis. Conditional power (the subject of most power analyses) may be distinguished from *actual* power, a parameter that depends on the (unknown) actual effect size. [It may seem strange to refer to the power of a test of H_0 as a parameter, because it is influenced not only by population parameters such as ψ_3 and σ_ε, but also by sample size. The power of a test is a function of the relevant *noncentrality parameter* (part of the definition of any noncentral distribution), which is also influenced by sample size.]

Power is influenced by all of the factors that influence precision, and it is also influenced by effect size. For any given sample size, the probability of rejecting a null hypothesis (or of obtaining a CI excluding zero) is greater (usually much greater) if the effect size is large than if it is small. This is not true of precision. For any given sample size, precision of estimation in dependent variable units is independent of effect size. The width of approximate standardized CIs is also independent of effect size. The width of noncentral (exact) standardized CIs does increase as effect size increases, but only slowly. Consider, for example, the change in precision and power that occurs when the difference between the two means in a two-group experiment is examined with $n = 20$ ($N = 40$), and $(\mu_1 - \mu_2)/\sigma_\varepsilon$ changes from zero to 1.0. This increase in effect size increases the power of a test of $H_0\colon \mu_1 - \mu_2$ from .05 (the lower limit of power when α = .05) to .87. The same increase in effect size increases the expected width of exact CIs on $(\mu_1 - \mu_2)/\sigma_\varepsilon$ by about 6%. That is, a change in effect size that produces a massive increase in power produces a small expected decrease in the precision of noncentral interval estimates of the population standardized mean difference.

The complement of power (β) is usually interpreted as an error rate (the Type II error rate) that has no analogue in CI analysis. Asymmetrical error rates (such as a Type I error rate of $\alpha = .05$ and a Type II error rate of $\beta = .20$) are usually recommended for power analysis (Cohen, 1988).

A detailed discussion of power analysis is beyond the scope of this book. We will, however, deal with the problem of determining the sample size required to control the probability of achieving confident direction inferences on contrasts, given a particular effect size.

Sample size and conditional power The sample size required to achieve conditional power of $(1-\beta)$ for a t- or F-based test of the null hypothesis $H_0\colon \psi_g = 0$ against the alternative hypothesis $H_1\colon |\psi_g| = \gamma\sigma_\varepsilon$ is approximately

$$n = 1 + \sum_j c_{gj}^2 \left(\frac{\mathrm{CC} + z_\beta}{\gamma} \right)^2 \tag{3.8}$$

where z_β is the $100(1-\beta)$th percentile point of the z distribution

γ is the standardized effect size specified by the alternative hypothesis

and CC is calculated on the assumption that the error variance is known.

If $\sum c^2 = 2$ (as is the case for comparisons), (3.8) applies to Tukey tests as well as t- and F-based (Scheffé) tests.

Tables F5 to F7 in Appendix F contain values of $\lambda = \mathrm{CC} + z_\beta$ for Tukey, Bonferroni-t and Scheffé tests for $\alpha = .05$ and .10 and power levels between .65 and .95. (For t tests controlling the PCER, use the first column of Table F5.) Having obtained the required value of λ from the relevant table, we can calculate the required sample size from

$$n = 1 + \sum_j c_{gj}^2 \left(\frac{\lambda}{\gamma} \right)^2 . \tag{3.9}$$

The n obtained from (3.9) is an approximation, because it makes use of the z distribution rather than relevant noncentral distributions. The approximation is usually a good one, however, particularly if non-integer values of n are always rounded up.

Consider the problem posed at the beginning of this section, where $\sum c^2 = 2$, $\gamma = 0.5$ and $(1-\beta) = .8$. From the first column in Table F5 we obtain $\lambda = 2.802$ when $(1-\beta) = .8$. From (3.9) we calculate

$$n = 1 + 2\left(\frac{2.802}{0.5} \right)^2 = 63.81.$$

Rounding up gives $n = 64$, so the overall sample size required is $Jn = 256$. (The same result is obtained from the exact method using a noncentral t distribution.)

Estimating actual power 'Observed power' statistics (estimates of actual power) are sometimes provided by statistical packages, presumably to provide a basis for some sort of inference when a test has failed to produce a statistically significant outcome. Point estimates of actual power are potentially misleading, however, because they encourage the researcher to treat the estimated effect size as a parameter rather than a statistic. Given a nonsignificant result (or a CI including zero), an estimate of the actual power of the test must be very imprecise, whatever the sample size. This can be demonstrated by constructing

CIs around power estimates, using a method outlined by Steiger and Fouladi (1997) and implemented by the *STATISTICA Power Analysis* program.

Suppose that we carry out a two-tailed t test on the difference between two independent means with a small sample (n = 10 observations per group), and we obtain a t ratio of 1.189 (p = .25). The exact 95% noncentral CI on the actual power of a .05 level test is (.050, .850). That is, the actual power of the test of $H_0: \mu_1 - \mu_2$ might be extremely low, very high or anywhere in between. Given the small sample size, it is not surprising that power is estimated so poorly. Suppose, however, that we obtain the same p value from such a test when the sample size is very large (n = 500 per group, t = 1.962, p = .25). The CI on actual power in this large-sample case is (.050, .875), which is as uninformative as in the small-sample case.

It should be noted that CIs on the standardized difference between means are very different in these two cases. As would be expected, the small-sample exact CI $[(\mu_1 - \mu_2)/\sigma_\varepsilon \in (-0.369, 2.101)]$ is relatively uninformative. On the other hand, the large-sample interval estimate (–0.051, 0.197) is very precise, justifying the inference that μ_1 and μ_2 are practically equivalent.

Why, then, is the large-sample CI on actual power so imprecise? One way to answer this question is in terms of the effect of sample size on the rate at which power increases as effect size increases. With a total sample size of N = 1000, the power of the test increases very rapidly as the standardized effect size increases from zero (when the power of the test is .05) to 0.197 (when the power of the test is .875). Another way to answer the question is to point out that an estimate of actual power provides no new information, given the outcome of the significance test (Hoenig and Heisey, 2001). If we know that both tests produce a p value of .25, then both estimates of actual power are redundant. It is of some interest to note that if $p = \alpha$, then the 100(1 – α)% CI on actual power is [α, (1 – α/2)]. That is, if we know from the test that p = .05, then we also know that the 95% CI on actual power must be (.05, .975), whatever the sample size. It is not clear whether the restatement of significance test outcomes in terms of actual power estimates is of any value in any context.[1] It is clear, however, that estimates of actual power cannot contribute to an experimenter's understanding of the reasons for a failure to reject a null hypothesis.

Precision or power?

In principle, power analysis can be useful at the point where experimenters are deciding on the sample size for an experiment, because it allows them to control the sensitivity of the experiment to nonzero effects of particular magnitudes. In practice, however, power analysis for this purpose often leads to the same conclusions as precision analysis: desirable sample sizes are likely to be prohibitively large.

Once an experiment has been run (perhaps with an overall sample size smaller than the experimenter might like), the information about precision provided by CIs is an integral part of CI inference. Given this information, it is difficult to see a legitimate additional role for power analysis. 'Observed power' statistics (estimates of actual power) are simply restatements of the outcomes of significance tests.

It is possible, of course, that a conditional power value (conditional on some nontrivial effect size) might throw some light on the reasons for a failure to establish the direction of a difference. For example, if $\mu_1 - \mu_2 = 0.3\sigma_\varepsilon$ and $n = 500$, the power of a .05-level t test is .997. Given this conditional power figure, it might be reasonable to interpret a nonsignificant t value as evidence that $|\mu_1 - \mu_2| < 0.3\sigma_\varepsilon$. This inference is redundant, however, given the CI $\mu_1 - \mu_2 \in (-0.048\sigma_\varepsilon, 0.200\sigma_\varepsilon)$. In general, post hoc power analysis cannot add anything to the information about precision provided by an appropriate CI.

Further reading

With the exception of Steiger and Fouladi (1997), there is very little accessible discussion of the precision of estimation of ANOVA model parameters or contrast values in the literature, primarily because of the dominant role of significance tests. Cohen (1988) provides a general treatment of power analysis, including an extensive discussion of the power of the ANOVA F test.

Questions and exercises

1. An experimenter wishes to control the maximum width of 90% CIs on contrasts by selecting an appropriate sample size for a three-group experiment. Use the relevant tables in Appendix F to determine the sample size required for the construction of approximate standardized CIs with a maximum half-width of $w = 0.25$ in the context of each of the following analyses:

(a) a planned contrasts analysis with individual CIs on two orthogonal contrasts (a {1, 1} contrast and a {2, 1} contrast);

(b) an analysis based on SCIs on all comparisons;

(c) a planned contrasts analysis based on SCIs on two comparisons;

(d) a post hoc analysis allowing for the construction of SCIs on any $\{m, r\}$ contrasts of interest to the experimenter after the data have been inspected.

2. An experiment with $J = 4$ groups is to be analysed on a post hoc basis with a set of 90% SCIs on any $\{m, r\}$ contrasts of interest. The experimenter intends to select a sample size that will produce a directional inference on a contrast with a probability of at least .75 if the effect is large ($\gamma = 0.8$).

(a) What procedure should be used and what sample size is required?

(b) What procedure and sample size would be required if the analysis were to be restricted to comparisons?

(c) What kind of trade-off is involved in choosing between the two analysis strategies [(a) and (b)]?

Note

1. Some statisticians define the concept of power in such a way that it can refer only to probabilities calculated before the experiment is run, or, at the very least, to probabilities calculated independently of the data. Given this restriction on the definition of power, it makes no sense to estimate actual power.

4 Simple Factorial Designs

In Chapter 2 we considered randomized experiments with a set of J qualitatively different experimental conditions or treatments, each of which can be regarded as a *level* of a single categorical independent variable. The Depression experiment, for example, includes $J = 4$ conditions, each of which is a level of a 'treatments for depression' variable. That variable is categorical, because the treatments are qualitatively (as distinct from quantitatively) different. In this and the following chapter we consider randomized experiments with two categorical independent variables (called *factors*). In a factorial design with two factors, each experimental condition is a combination of a level of the first factor and a level of the second factor. The two factors are *crossed*, meaning that each of the J levels of the first factor is combined with each of the K levels of the second factor to produce a set of $J \times K$ experimental conditions.

In this chapter we consider CI analyses of *factorial effects* (simple effects, main effects and interaction effects) defined in the context of various models of data from a simple 2×2 factorial design, where each factor has only two levels. Suppose that an experimenter wishes to examine the effects of two factors on driving performance: sleep deprivation, and the level of NVH (noise, vibration and harshness) produced by the vehicle. Sleep deprivation is the factor of primary interest. The second factor is included to see whether the magnitude of the sleep deprivation effect varies across NVH levels. The sleep deprivation factor, which we will call Factor A, has $J = 2$ levels: a_1 (12 hours of sleep deprivation) and a_2 (no sleep deprivation). The NVH factor (Factor B) has $K = 2$ levels: b_1 (high NVH) and b_2 (low NVH). The design has $J \times K = 4$ experimental conditions or *cells*:

a_1b_1: 12 hours of sleep deprivation, high NVH
a_1b_2: 12 hours of sleep deprivation, low NVH
a_2b_1: no sleep deprivation, high NVH
a_2b_2: no sleep deprivation, low NVH.

After the appropriate number of hours of sleep deprivation (12 for subjects in a_1 cells, zero for subjects in a_2 cells), subjects are tested in a driving simulator that allows for manipulation of NVH levels. The dependent variable is an error score, a relatively high score indicating relatively poor performance.

Factorial effect contrasts defined on cell means

The *cell means* model is

$$Y_{ijk} = \mu_{jk} + \varepsilon_{ijk} \tag{4.1}$$

where Y_{ijk} is the score on the dependent variable Y obtained by subjects in cell jk ($j = 1, 2, \ldots, J$; $k = 1, 2, \ldots, K$)

μ_{jk} is the population mean of cell jk (that is, the expected value of Y_{ijk})

ε_{ijk} is the value of subject i in cell jk on an error variable ε with a mean of zero within treatment population jk.

The cell means model is closely related to the means model (2.2) discussed in Chapter 2. The two models share the same assumptions about error distributions, and both models allow for the definition of contrasts on means. It will be convenient to introduce some of the basic ideas about factorial effects by considering factorial contrasts on cell means, mainly because some factorial effects (specifically simple effects) are not defined by the factorial ANOVA models we will consider later.

Although a large number of different contrasts (including 25 different $\{m, r\}$ contrasts) can be defined on the means of the four cells of a 2×2 factorial design, only seven are *factorial effect* contrasts. These contrasts are defined in Table 4.1, where the coefficients c_{jk} ($j = 1, 2$; $k = 1, 2$) refer to the cell means μ_{jk}. The first of the two subscripts refers to a level of Factor A (Sleep deprivation) and the second subscript refers to a level of Factor B (NVH).

The contrast $\psi_{A(b_1)} = \mu_{11} - \mu_{21}$ is the *simple effect* of Factor A (the difference between 12 and 0 hours of sleep deprivation) at the first level of Factor B

Table 4.1 *Simple, main and interaction effect contrasts on cell means in a 2×2 design*

Contrast	Type of effect	c_{11}	c_{12}	c_{21}	c_{22}	$\sum_j \sum_k c^{+}_{jk}$	$\sum_j \sum_k c^2_{jk}$
$\psi_{A(b_1)}$	Simple	1	0	–1	0	1	2
$\psi_{A(b_2)}$	Simple	0	1	0	–1	1	2
$\psi_{B(a_1)}$	Simple	1	–1	0	0	1	2
$\psi_{B(a_2)}$	Simple	0	0	1	–1	1	2
ψ_A	Main	0.5	0.5	–0.5	–0.5	1	1
ψ_B	Main	0.5	–0.5	0.5	–0.5	1	1
ψ_{AB}	Interaction	1	–1	–1	1	2	4

(driving a vehicle with a high NVH level). There are two A simple effect contrasts ($\psi_{A(b_1)}$ and $\psi_{A(b_2)}$) because Factor B has two levels. Similarly, it is possible to define two B simple effect contrasts, the first of which ($\psi_{B(a_1)}$) is the NVH effect (the difference between high and low NVH levels) on drivers with 12 hours of sleep deprivation, while the second ($\psi_{B(a_2)}$) is the NVH effect on drivers who are not sleep deprived. All four simple effect contrasts in a 2×2 design are comparisons on cell means.

Contrasts ψ_A and ψ_B are *main effect* contrasts. In a factorial experiment a main effect contrast is a contrast on levels of one factor, averaged across levels of all other factors. In this case, the A main effect contrast is concerned with the difference between 12 and 0 hours of sleep deprivation, averaged across NVH levels, while the B main effect contrast is concerned with the difference between high and low NVH levels, averaged across the two levels of sleep deprivation. All of the simple effect and both of the main effect contrasts are mean difference contrasts.

The *AB interaction effect* contrast is not a mean difference contrast. The interaction contrast can be interpreted in at least two ways. First, it is the difference between the magnitudes of the two A simple effect contrasts. That is,

$$\psi_{A(b_1)} - \psi_{A(b_2)} = (\mu_{11} - \mu_{21}) - (\mu_{12} - \mu_{22}) = \mu_{11} - \mu_{12} - \mu_{21} + \mu_{22} = \psi_{AB}.$$

Second, it is the difference between the magnitudes of the two B simple effect contrasts. That is,

$$\psi_{B(a_1)} - \psi_{B(a_2)} = (\mu_{11} - \mu_{12}) - (\mu_{21} - \mu_{22}) = \mu_{11} - \mu_{12} - \mu_{21} + \mu_{22} = \psi_{AB}.$$

The scaling of the coefficients of the interaction contrast is determined by these interpretations. Any interaction contrast in a two-factor design must have coefficients scaled so that $\sum c_{jk}^{+} = 2.0$ if it is to be interpreted as a mean difference (between levels of one factor) in the magnitude of a mean difference (between levels of the other factor). The entries in the final column of Table 4.1 imply that if cell sample sizes are equal, the standard error of the interaction contrast is larger than that of simple effect contrasts, which in turn is larger than that of main effect contrasts.

If the interaction contrast has a population value of zero (an important special case), then the magnitude of each A simple effect contrast is equal to the magnitude of the A main effect contrast, and the magnitude of each B simple effect contrast is equal to the magnitude of the B main effect contrast. Suppose, for example, that $\mu_{11} = 47$, $\mu_{12} = 43$, $\mu_{21} = 37$ and $\mu_{22} = 33$. Then $\psi_{AB} = 0$, $\psi_{A(b_1)} = \psi_{A(b_2)} = \psi_A = 10$, and $\psi_{B(a_1)} = \psi_{B(a_2)} = \psi_B = 4$. If there is no interaction, then the magnitude of the sleep deprivation effect does not vary across the two NVH levels, so the most precise estimate of that effect is the estimate of the A main effect contrast. This follows from the fact that the A main effect contrast has a smaller value of $\sum c^2$ than do the A simple effect contrasts. Similarly, the absence of interaction between the two factors implies that there is only one NVH effect to be estimated, and the best way to estimate that effect is to estimate the value of the *B* main effect contrast, ignoring the individual *B* simple effect contrasts.

The two-factor main effects model

While the population value of the interaction contrast is unlikely to be exactly zero, it may well be so close to zero that the two sleep deprivation simple effects (and the two NVH simple effects) are practically equivalent. If so, then all of the nontrivial variation between population cell means can be described by a *main effects model*:

$$\mathrm{E}(Y_{ijk}) = \mu + \alpha_j + \beta_k \tag{4.2}$$

where Y_{ijk} is the score on the dependent variable Y obtained by subject i ($i = 1, 2, \ldots, n_{jk}$) in cell jk

α_j is a main effect parameter for Factor A $(\sum_j \alpha_j = 0)$

and β_k is a main effect parameter for Factor B $(\sum_k \beta_k = 0)$.

Suppose that $\alpha_1 = 5$ (implying that $\alpha_2 = -5$, because of the zero-sum constraint on the α_j parameters) and $\beta_1 = 2$ (implying that $\beta_2 = -2$, because of the constraint on the β_k parameters). If $\mu = 40$, then it follows from (4.2) that

$$\mu_{11} = 40 + 5 + 2 = 47$$
$$\mu_{12} = 40 + 5 - 2 = 43$$
$$\mu_{21} = 40 - 5 + 2 = 37$$

and $$\mu_{22} = 40 - 5 - 2 = 33.$$

The four cell means are presented graphically in Figure 4.1. The slope of the upper line joining the two means at the first level of Factor *A* shows that $\mu_{11} > \mu_{12}$, implying that $\psi_{B(a_1)} > 0$. Similarly, the lower line implies that $\psi_{B(a_2)} > 0$. The fact that each of the endpoints of the upper line is higher than the corresponding endpoint of the lower line implies that each of the *A* simple effect contrasts has a positive value. The fact that the average of the means joined by the upper line is higher than the average of the means joined by the lower line implies that the *A* main effect contrast is positive, and the fact that the average of the b_1 means is higher than the average of the b_2 means implies that the *B* main effect contrast is also positive. The size of the sleep deprivation effect, defined as the difference between the α_j parameters at 12 and 0 hours of sleep deprivation, is $\alpha_1 - \alpha_2 = 5 - (-5) = 10$, the value of the *A* simple and main effect contrasts. Similarly, the size of the NVH effect is $\beta_1 - \beta_2 = 2 - (-2) = 4$, the value of the *B* simple and main effect contrasts.

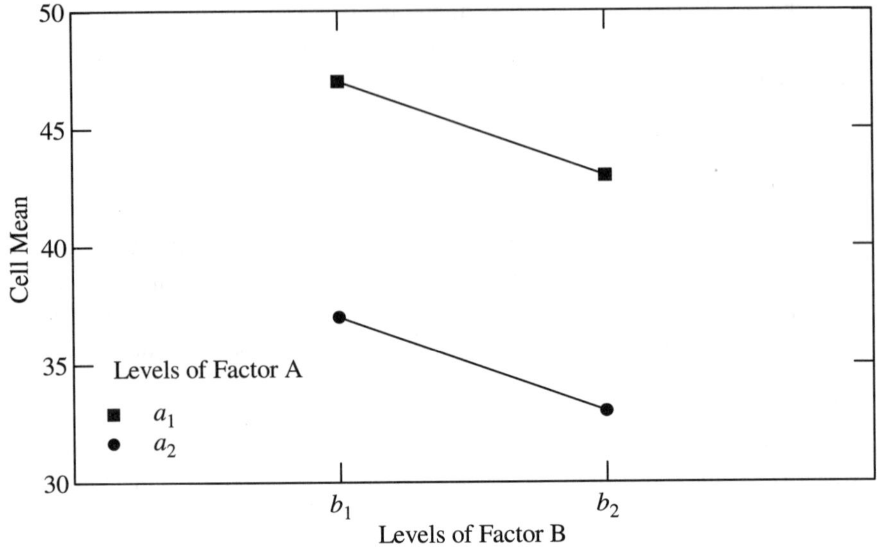

Figure 4.1 *Population means in a 2 ×2 design when effects are additive.*

The main effects model implies that the magnitude of the effect of each factor does not vary across levels of the other factor, and that the combined effect of the two factors is as an *additive* combination of the individual effects. When the sleep deprivation effect (10) is combined with the NVH effect (4), the combined effect is simply the sum of the individual effects (10 + 4 = 14), as can be seen by comparing μ_{11} with μ_{22}. Additivity of sleep deprivation and NVH effects is represented in Figure 4.1 by the *parallelism* of the lines representing the two *B* simple effects contrasts, implying that these contrasts have identical values, so that the value of the *AB* interaction contrast is zero.

As the equation defining the main effects model (4.2) suggests, main effects are the only effects requiring estimation when there is no interaction; simple effects are redundant, given the main effect parameters.

Interaction If the magnitudes of the simple effects of one factor vary nontrivially across levels of the other factor (that is, if the interaction is not trivially small), then the main effects model provides an oversimplified and possibly misleading account of the pattern of differences between population means. Suppose now that $\mu_{11} = 44$, $\mu_{12} = 46$, $\mu_{21} = 40$ and $\mu_{22} = 30$. This pattern of means is shown in Figure 4.2, and the magnitudes of the various effects (simple, main and interaction) are shown in Table 4.2. The sleep deprivation effect is much greater when the NVH level is low ($\psi_{A(b_2)} = 16$) than when it is high ($\psi_{A(b_1)} = 4$), and the NVH effect on sleep-deprived drivers ($\psi_{B(a_1)} = -2$) differs in sign as well as magnitude from the effect on drivers who

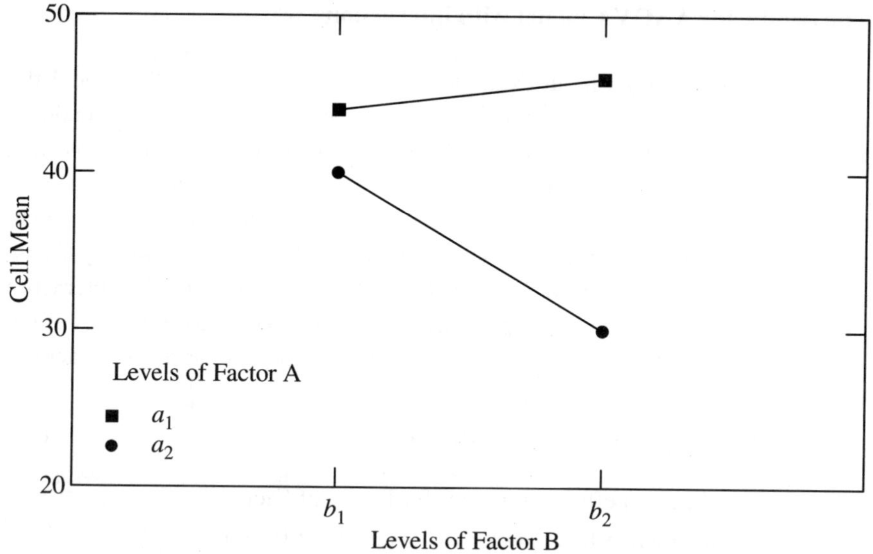

Figure 4.2 *Population means in a 2 × 2 design when effects are not additive*

are not sleep deprived ($\psi_{B(a_2)} = 10$). Given the relatively large value of the interaction contrast ($\psi_{AB} = -12$), the information provided by the main effect contrasts ($\psi_A = 10$ and $\psi_B = 4$) is at best incomplete, and at worst misleading.

One way of dealing with the deficiencies of the main effects model in the presence of heterogeneous simple effects is to adopt an alternative model that includes simple effect parameters. This is not the approach usually taken, although it does have some advantages over the extended two-factor model that is usually adopted. We will examine simple effect models later in this chapter.

Table 4.2 *Values of simple, main and interaction effect contrasts in a 2 × 2 design when* $\mu_{11} = 44$, $\mu_{12} = 46$, $\mu_{21} = 40$ *and* $\mu_{22} = 30$

Contrast	Type of effect	c_{11}	c_{12}	c_{21}	c_{22}	Value
$\psi_{A(b_1)}$	Simple	1	0	–1	0	4
$\psi_{A(b_2)}$	Simple	0	1	0	–1	16
$\psi_{B(a_1)}$	Simple	1	–1	0	0	–2
$\psi_{B(a_2)}$	Simple	0	0	1	–1	10
ψ_A	Main	0.5	0.5	–0.5	–0.5	10
ψ_B	Main	0.5	–0.5	0.5	–0.5	4
ψ_{AB}	Interaction	1	–1	–1	1	–12

The two-factor ANOVA model with interaction

The main effects model is an example of an *unsaturated* ANOVA model that does not necessarily provide a perfect fit to the pattern of means it is intended to account for or 'explain'. In the 2 × 2 case, the main effects model has three independent parameters (μ, α_1 and β_1), the remaining parameters being made redundant by zero-sum constraints. Because there are fewer independent parameters (three) than means (four), the model does not necessarily fit the means perfectly. The *saturated* two-factor ANOVA model includes interaction parameters, and these additional parameters allow the model to fit any set of cell means perfectly. The saturated two-factor ANOVA model for a $J \times K$ factorial design is

$$Y_{ijk} = \mu + \alpha_j + \beta_k + \alpha\beta_{jk} + \varepsilon_{ijk} \tag{4.3}$$

where α_j is the main effect parameter for level j of Factor A

β_k is the main effect parameter for level k of Factor B

$\alpha\beta_{jk}$ is the AB interaction parameter for cell jk

and ε_{ijk} is the error component associated with Y_{ijk}.

The model is seriously overparameterized (nine parameters are required to fit the four cell means of a 2 × 2 design), so the following constraints are usually imposed to reduce the number of independent parameters to the number of cells:

$$\sum_j \alpha_j = 0,$$

$$\sum_k \beta_k = 0,$$

$$\sum_j \alpha\beta_{jk} = 0 \text{ for each of the } K \text{ levels of Factor } B$$

and $$\sum_k \alpha\beta_{jk} = 0 \text{ for each of the } J \text{ levels of Factor } A.$$ [1]

Accounting for variation between cell means The meaning of the parameters of the model can be seen if the cell means from the sleep-deprivation × NVH experiment are presented in a matrix where the rows are levels of Factor A and the columns are levels of Factor B. Cell, row and column means are

	b_1	b_2	Row means
a_1	$\mu_{11} = 44$	$\mu_{12} = 46$	$\mu_{1.} = 45$
a_2	$\mu_{21} = 40$	$\mu_{22} = 30$	$\mu_{2.} = 35$
Column means	$\mu_{.1} = 42$	$\mu_{.2} = 38$	$\mu = 40$

The first of the subscripts attached to a mean refers to a level of Factor *A*, the second to a level of Factor *B*. The use of a dot in place of a subscript (as in $\mu_{1.} = 45$) indicates that the mean averages across rows or columns: $\mu_{1.}$ is the average of the means in row 1, averaged across columns. Dot notation will be used when it is necessary to do so to avoid ambiguity ($\mu_{1.}$ is not the same as $\mu_{.1}$). If there is no risk of ambiguity, dots will sometimes be dropped ($\mu = \mu_{..}$).

The *A* main effect parameters are

$$\alpha_1 = \mu_{1.} - \mu = 45 - 40 = 5$$

and

$$\alpha_2 = \mu_{2.} - \mu = 35 - 40 = -5.$$

Thus *A* main effect parameters account for variation between rows (levels of Factor *A*) averaged across columns (levels of Factor *B*). The contributions of *A* main effect parameters to cell, row and column means are

	b_1	b_2	Row means
a_1	$\alpha_1 = 5$	$\alpha_1 = 5$	5
a_2	$\alpha_2 = -5$	$\alpha_2 = -5$	–5
Column means	0	0	0

Similarly, *B* main effect parameters contribute only to variation between columns (levels of Factor *B*) averaged across rows (levels of Factor *A*). The values of these parameters are

$$\beta_1 = \mu_{.1} - \mu = 42 - 40 = 2$$

and

$$\beta_2 = \mu_{.2} - \mu = 38 - 40 = -2,$$

and their contributions to cell means are:

	b_1	b_2	Row means
a_1	$\beta_1 = 2$	$\beta_2 = -2$	0
a_2	$\beta_1 = 2$	$\beta_2 = -2$	0
Column means	2	–2	0

Interaction parameters account for all of the variation between cell means that remains when the additive combination of main effect parameters is removed. That is, $\alpha\beta_{jk} = (\mu_{jk} - \mu) - (\alpha_j + \beta_k)$. The interaction parameters are

$$\alpha\beta_{11} = \mu_{11} - \mu - \alpha_1 - \beta_1 = 44 - 40 - 5 - 2 = -3$$

$$\alpha\beta_{12} = \mu_{12} - \mu - \alpha_1 - \beta_2 = 46 - 40 - 5 - (-2) = 3$$

$$\alpha\beta_{21} = \mu_{21} - \mu - \alpha_2 - \beta_1 = 40 - 40 - (-5) - 2 = 3$$

and

$$\alpha\beta_{22} = \mu_{22} - \mu - \alpha_2 - \beta_2 = 30 - 40 - (-5) - (-2) = -3.$$

As the following matrix of interaction parameters shows, interaction makes no contribution to variation between rows (averaged across columns), or to variation between columns (averaged across rows).

	b_1	b_2	Row means
a_1	$\alpha\beta_{11} = -3$	$\alpha\beta_{12} = 3$	0
a_2	$\alpha\beta_{21} = 3$	$\alpha\beta_{22} = -3$	0
Column means	0	0	0

The matrix of interaction parameters is *doubly centred*, meaning that the parameters sum to zero in each row and also in each column. (The matrix of A main effect parameters is column centred, and the matrix of B main effect parameters is row centred).

Effect size parameters An analysis based on the two-factor ANOVA model is essentially a collection of three sub-analyses, each of which provides inferences about one of the three types of effect parameter (A main effect, B main effect, and AB interaction effect). Each sub-analysis can produce estimates of the values of the following functions of one type of parameter:

- Cohen's f, indicating the degree of heterogeneity among the relevant effect parameters;
- raw and standardized contrasts (generalizations of Cohen's d) on the relevant effect parameters.

The three f parameters are

$$f_A = \sqrt{\frac{\sum_j \alpha_j^2}{J\sigma_\varepsilon^2}}, \quad f_B = \sqrt{\frac{\sum_k \beta_k^2}{K\sigma_\varepsilon^2}} \quad \text{and} \quad f_{AB} = \sqrt{\frac{\sum_j\sum_k \alpha\beta_{jk}^2}{JK\sigma_\varepsilon^2}},$$

each of which can be interpreted as the 'standard deviation' of the relevant type of effect parameter. These effect size indices have the same advantages and disadvantages as the f index defined on the effect parameters of the single-factor ANOVA model. They have not been widely used, except for the purpose of power analysis (Cohen, 1988).

If f_{AB} is trivially small, then the A and B effects are approximately additive, and the relatively simple main effects model (4.2) will fit the data almost as well as the saturated model including interaction parameters (4.3).

Sources of variation in a balanced two-factor design

A *balanced* two-factor design provides the basis for a partition of between-cells variation into three orthogonal (statistically independent) additive components, each of which provides the basis for inferences about one of the three types of effect parameters defined by the two-factor ANOVA model. Any factorial experiment with equal cell frequencies ($n_{jk} = n$ in the case of a two-factor design) is balanced. All of the discussion of factorial designs in this chapter will be based on the assumption that sample sizes are equal. The CI analyses discussed in this chapter can be applied to data from unbalanced (or nonorthogonal) designs, provided that the analyses are based on saturated models [such as (4.1) and (4.3)] rather than unsaturated models such as the main effects model (4.2).

The two-factor model can be written in terms of deviations from population means as

$$\begin{aligned} Y_{ijk} - \mu &= \alpha_j + \beta_k + \alpha\beta_{jk} + \varepsilon_{ijk} \\ &= (\mu_j - \mu) + (\mu_k - \mu) + (\mu_{jk} - \mu_j - \mu_k + \mu) + (Y_{ijk} - \mu_{jk}) \end{aligned}$$

where μ_j is the average of the *K* cell means at level *j* of Factor *A*

and μ_k is the average of the *J* cell means at level *k* of Factor *B*.

The analogous partition of variability in the data from a balanced two-factor experiment is

$$Y_{ijk} - M = (M_j - M) + (M_k - M) + (M_{jk} - M_j - M_k + M) + (Y_{ijk} - M_{jk}).$$

The within-cells component $(Y_{ijk} - M_{jk})$ is identical to the within-groups component in the simpler single-factor model, and in both cases is the basis for the error term MS_E. The two-factor ANOVA procedure partitions variation between cell (group) means $(M_{jk} - M)$ into three components:

- $(M_j - M)$, the basis for the *SS* between rows of the $J \times K$ matrix of sample means,
- $(M_k - M)$, the basis for the *SS* between columns in that matrix, and
- $(M_{jk} - M_j - M_k + M)$, the basis for residual variation between cells.

To see how this partition works in the 2 × 2 case, we will examine a data set with $n = 25$ subjects per cell produced by a simulation of the Sleep deprivation × NVH experiment where the population means are those given for the saturated model and $\sigma_\varepsilon^2 = 25$. (This data set is in the *PSY* input file *SD×NVH.in*.) The simulation produced the sample means shown in Table 4.3. As would be expected, the pattern of sample cell means (shown in Figure 4.3) is similar to the pattern of population cell means (shown in Figure 4.2).

Variation between cells A standard two-way ANOVA partitions deviation means $(M_{jk} - M)$ into row, column and residual components. Row deviation means $(M_j - M)$ are

$$(M_{1.} - M) = 44.46 - 38.47 = 5.99$$

and

$$(M_{2.} - M) = 32.48 - 38.47 = -5.99.$$

The *sum of squares between rows* is the sum of squared deviations of the J row deviation means multiplied by the number of subjects contributing to each row mean (Kn). We will refer to the between-rows SS as SS_A:

$$SS_A = Kn\sum_j (M_j - M)^2 = 2 \times 25[(5.99)^2 + (-5.99)^2] = 3588.01.$$

Row deviation means necessarily sum to zero, so the number of degrees of freedom for variation between rows is $(J - 1)$.

The mean square between rows is obtained by dividing SS_A by ν_A:

$$MS_A = \frac{SS_A}{\nu_A} = \frac{3588.01}{1} = 3588.01.$$

The expected value of MS_A is

$$\mathrm{E}(MS_A) = \sigma_\varepsilon^2 + \frac{Kn\sum_j \alpha_j^2}{J-1},$$

so the magnitude of this mean square is influenced by the magnitude of the A main effect parameters, but not by any of the other effect parameters defined by the two-factor ANOVA model.

Column deviation means $(M_k - M)$ are

$$(M_{.1} - M) = 40.12 - 38.47 = 1.65$$

and

$$(M_{.2} - M) = 36.82 - 38.47 = -1.65.$$

Table 4.3 *Means (and standard deviations) from Sleep deprivation* $\times$ *NVH* data set

	b_1 High NVH	b_2 Low NVH	M_j
a_1 Sleep Dep	43.44 (4.59)	45.48 (5.16)	44.46
a_2 No Sleep Dep	36.80 (5.30)	28.16 (4.06)	32.48
M_k	40.12	36.82	$M = 38.47$

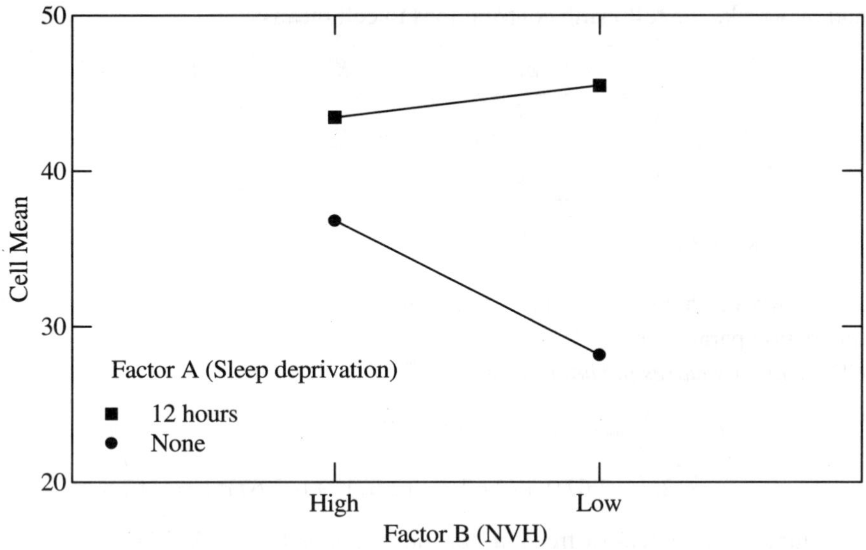

Figure 4.3 *Profiles of means from Sleep deprivation × NVH data set*

The *sum of squares between columns* is the sum of squared deviations of the K column deviation means multiplied by the number of subjects contributing to each column mean (Jn). The sum of squares between columns is

$$SS_B = Jn\sum_k (M_k - M)^2 = 2 \times 25\ [(1.65)^2 + (-1.65)^2] = 272.25.$$

The number of degrees of freedom for variation between columns is $(K - 1)$. The mean square between columns is

$$MS_B = \frac{SS_B}{\nu_B} = \frac{272.25}{1} = 272.25$$

and the expected value of MS_B is

$$\mathrm{E}(MS_B) = \sigma_\varepsilon^2 + \frac{Jn\sum_j \beta_j^2}{K-1}.$$

Subtracting row and column deviation means from the cell deviation means gives the *interaction means*

$$(M_{jk} - M) - (M_j - M) - (M_k - M) = (M_{jk} - M_j - M_k + M).$$

The interaction means in this case are

$$(M_{11} - M_{1.} - M_{.1} + M) = 43.44 - 44.46 - 40.12 + 38.47 = -2.67$$
$$(M_{12} - M_{1.} - M_{.2} + M) = 45.48 - 44.46 - 36.82 + 38.47 = 2.67$$
$$(M_{21} - M_{2.} - M_{.1} + M) = 36.80 - 32.48 - 40.12 + 38.47 = 2.67$$
$$(M_{22} - M_{2.} - M_{.2} + M) = 28.16 - 32.48 - 36.82 + 38.47 = -2.67$$

and they make the following contributions to cell means:

	b_1	b_2	Row means
a_1	–2.67	2.67	0
a_2	2.67	–2.67	0
Column means	0	0	0

The matrix of interaction means is necessarily doubly centred, as is the matrix of interaction parameters.

The *sum of squares for interaction* is

$$\begin{aligned} SS_{AB} &= n\sum_j\sum_k(M_{jk} - M_j - M_k + M)^2 \\ &= 25[(-2.67)^2 + (2.67)^2 + (2.67)^2 + (-2.67)^2] = 712.89. \end{aligned}$$

The number of degrees of freedom for interaction is $(J - 1)(K - 1)$, the product of the number of degrees of freedom between rows and the number of degrees of freedom between columns. In the 2×2 case $(J - 1)(K - 1) = 1$. If we refer to the data to calculate any one of the four interaction means, such as $(M_{11} - M_{1.} - M_{.1} + M) = -2.67$, the zero-sum constraints allow us to determine the remaining three interaction means without further reference to the data.

Mean square interaction is

$$MS_{AB} = \frac{SS_{AB}}{\nu_{AB}} = \frac{712.89}{1} = 712.89$$

with an expected value of

$$\mathrm{E}(MS_{AB}) = \sigma_\varepsilon^2 + \frac{n\sum_j\sum_k \alpha\beta_{jk}^2}{(J-1)(K-1)}.$$

Variation within cells The sum of squares within cells is calculated in the same way as SS_W in a one-way ANOVA. In this case SS_W (or SS_E) is 2213.76. In a two-factor design there are $(N - JK)$ [$= JK(n - 1)$ when cell frequencies are equal] degrees of freedom for variation within cells, so

$$MS_E = 2213.76/(4 \times 24) = 23.06.$$

The expected value of MS_E is σ_ε^2.

Heterogeneity inference

The partition of variation is summarized in Table 4.4, which also contains the F statistics (and associated p values) that provide the basis for a heterogeneity inference within each of the three sub-analyses. Each F statistic (F_A, F_B and F_{AB}) is a variance ratio obtained by dividing the relevant mean square (MS_A, MS_B or MS_{AB}) by MS_E. Given the ANOVA-model assumptions about error distributions (see Chapter 2), the F statistics are distributed as noncentral F distributions with degrees of freedom and noncentrality parameters shown in Table 4.5.

It is possible to use the three F statistics to construct noncentral CIs on the heterogeneity indices f_A, f_B and f_{AB}. There is little point in doing so for an effect with only 1 degree of freedom, however, because in that case the CI on f is redundant, given the CI on the corresponding standardized contrast. We will consider CIs on f parameters in factorial designs in Chapter 5, which deals with designs where each effect has multiple degrees of freedom.

The p values in the final column of Table 4.4 indicate that F_A, F_B and F_{AB} are statistically significant at any conventional α level, from which it may be concluded that the homogeneity hypotheses

$$H_A: \ \alpha_1 = \alpha_2 = 0$$

$$H_B: \ \beta_1 = \beta_2 = 0$$

and $$H_{AB}: \alpha\beta_{11} = \alpha\beta_{12} = \alpha\beta_{21} = \alpha\beta_{22} = 0$$

are all false. These F tests (with $\nu_1 = 1$) are redundant if CIs are to be constructed on main effect and interaction contrasts.

Contrasts on parameters of the two-factor ANOVA model

The two-factor ANOVA model for a 2 × 2 design has two A main effect parameters (α_1 and α_2) so the only possible mean difference contrast with coefficients referring to those parameters has a coefficient vector of $\mathbf{c}'_A$ = [1 –1] (or $-\mathbf{c}'_A$ = [–1 1]):

$$\psi_A = \mathbf{c}'_A\alpha = \alpha_1 - \alpha_2 = (\mu_{1.} - \mu) - (\mu_{2.} - \mu) = \mu_{1.} - \mu_{2.}.$$

This is the same contrast as the A main effect contrast defined in Table 4.1,

Table 4.4 *Two-way ANOVA summary table*

Source	*SS*	*df*	*MS*	*F*	*p*
A (Sleep Dep)	3588.01	1	3588.01	155.59	< .001
B (NVH)	272.25	1	272.25	11.81	< .001
AB	712.89	1	712.89	30.91	< .001
Error	2213.76	96	23.06		
Total	6786.91	99			

Table 4.5 *Parameters of noncentral F distributions assumed to generate F statistics in two-way ANOVA*

F statistic	Parameters of noncentral *F* distribution		
	ν_1	ν_2	δ
F_A	$(J-1)$	$(N-JK)$	$\dfrac{Kn\sum_j \alpha_j^2}{\sigma_\varepsilon^2}$
F_B	$(K-1)$	$(N-JK)$	$\dfrac{Jn\sum_k \beta_k^2}{\sigma_\varepsilon^2}$
F_{AB}	$(J-1)(K-1)$	$(N-JK)$	$\dfrac{n\sum_j\sum_k \alpha\beta_{jk}^2}{\sigma_\varepsilon^2}$

with two coefficients referring to α_j parameters rather than with four coefficients referring to cell means. Similarly, the only possible *B* main effect mean difference contrast with coefficients referring to β_k parameters is

$$\psi_B = \mathbf{c}'_B\beta = \beta_1 - \beta_2 = (\mu_{.1}-\mu)-(\mu_{.2}-\mu) = \mu_{.1}-\mu_{.2}$$

with coefficient vector $\mathbf{c}'_B = [1\ \ -1]$.

Unlike main effect parameters, which refer to factor levels, each interaction parameter $\alpha\beta_{jk}$ refers to a combination of factor levels, that is to a cell. Any contrast on cell means is also a contrast on interaction parameters if

$$\sum_j c_{jk} = 0 \text{ for each } k \text{ (each level of Factor } B)$$

and $$\sum_k c_{jk} = 0 \text{ for each } j \text{ (each level of Factor } A).$$

If the coefficients of the *AB* contrast defined at the beginning of this chapter are arranged in a matrix with rows as levels of Factor *A* and columns as levels of Factor *B*, we can see that this contrast does indeed satisfy these constraints. The coefficient matrix is

$$\mathbf{C}_{AB} = \begin{bmatrix} c_{11} & c_{12} \\ c_{21} & c_{22} \end{bmatrix} = \begin{bmatrix} 1.0 & -1.0 \\ -1.0 & 1.0 \end{bmatrix}.$$

The coefficients sum to zero in each row and in each column so $\mathbf{C}_{AB}$ is a doubly centred coefficient matrix. As a consequence,

$$\sum_j\sum_k c_{jk}\mu_{jk} = \sum_j\sum_k c_{jk}\alpha\beta_{jk} = \psi_{AB}.$$

When there is only 1 degree of freedom for an effect in a two-way ANOVA model (as is the case for all three effects in a 2×2 design), the sum of squares for the only contrast within that effect is equal to the sum of squares for the effect. That is, $SS(\hat{\psi}_A) = SS_A$, $SS(\hat{\psi}_B) = SS_B$ and $SS(\hat{\psi}_{AB}) = SS_{AB}$. In the special case of a 2×2 design, the standard two-factor ANOVA procedure is therefore equivalent to a planned contrasts analysis.

Confidence intervals on main effect and interaction contrasts We have already seen that the standard analysis of the Sleep deprivation × NVH data set rejects all three homogeneity hypotheses, implying that the associated planned contrasts analysis would produce inequality inferences on all three contrasts. A planned analysis produces the following individual CIs:

Contrast	$\hat{\psi}_g$	$\hat{\sigma}_{\hat{\psi}_g}$	95% confidence interval
ψ_A	11.980	0.960	$\psi_A \in (10.074,\ 13.886)$
ψ_B	3.300	0.960	$\psi_B \in (1.394,\ 5.206)$
ψ_{AB}	–10.680	1.921	$\psi_{AB} \in (-14.493,\ -6.867)$.

When interpreting these intervals we will suppose that a difference of about 2.0 is the smallest difference considered to be important. The main effect CIs indicate that 12 hours of sleep deprivation has a very large detrimental effect on performance in the driving simulator task (averaged across noise, vibration and harshness levels), and that that a high level of NVH produces more errors than a low NVH level (averaged across sleep deprivation levels). It is not clear whether the NVH main effect is large enough to be of any real concern. It is clear from the CI on the interaction contrast, however, that the difference between the size of the NVH effect on sleep-deprived subjects and the effect on those not sleep deprived is large and negative.

The most obvious limitation of this analysis is the absence of inferences on simple effects contrasts, due to the fact that the two-factor ANOVA model does not include simple effect parameters.

A simple effects model

To see what a model including simple effects might look like, consider the following parameters:

$$\begin{aligned} \alpha_j(b_k) &= \mu_{jk} - \mu_k \\ &= (\mu_j - \mu) + (\mu_{jk} - \mu_j - \mu_k + \mu) \\ &= \alpha_j + \alpha\beta_{jk} \end{aligned} \tag{4.4}$$

and

$$\begin{aligned} \beta_k(a_j) &= \mu_{jk} - \mu_j \\ &= (\mu_k - \mu) + (\mu_{jk} - \mu_j - \mu_k + \mu) \\ &= \beta_k + \alpha\beta_{jk}. \end{aligned} \tag{4.5}$$

The parameter $\alpha_j(b_k)$ is the simple effect parameter for level j of Factor A at level k of Factor B. The fact that $\alpha_j(b_k)$ can be expressed as a linear combination of two parameters from the two-factor model does not mean that a simple effect parameter is more complex than its name suggests; it means that the two-factor model does not provide a simple account of simple effects. [A main effect parameter can be made to look complicated by defining it in terms of simple and interaction effects: $\alpha_j = \alpha_j(b_k) - \alpha\beta_{jk}$.] The parameter $\beta_k(a_j)$ is the simple effect parameter for treatment k of Factor B at level j of Factor A.

Because the complete set of A simple effect parameters accounts for all of the variation between cell means except for that accounted for by the B main effect, one alternative to the two-factor ANOVA model is the *A simple effects model*:

$$Y_{ijk} = \mu + \alpha_j(b_k) + \beta_k + \varepsilon_{ijk}. \tag{4.6}$$

The B main effect parameter β_k is included in order to ensure that (4.6) is a saturated model accounting for all of the variation between cell means. Note that the A simple effects model makes no reference to B simple effects.

The A simple effects model can be written in terms of deviations from population means as

$$Y_{ijk} - \mu = (\mu_{jk} - \mu_k) + (\mu_k - \mu) + (Y_{ijk} - \mu_{jk}).$$

The analogous partition of variability in the data from a two-factor experiment is

$$Y_{ijk} - M = (M_{jk} - M_k) + (M_k - M) + (Y_{ijk} - M_{jk})$$

where the $(M_{jk} - M_k)$ components are *A simple effect means*. The A simple effect means from the current data set are

	b_1	b_2	Row means
a_1	3.32	8.66	5.99
a_2	–3.32	–8.66	–5.99
Column means	0	0	0

Because only variation between column means (averaged across rows) has been removed, everything orthogonal to (statistically independent of) that source remains. Thus variation between the A simple effect means can be used to define a family of all factorial contrasts involving any difference between levels of Factor A: the two simple effect contrasts $\psi_{A(b_1)}$ and $\psi_{A(b_2)}$, the main effect contrast ψ_A and the interaction contrast ψ_{AB}. We will refer to this family of contrasts as $A(B)$. All contrasts in the family refer to the difference between levels of Factor A on some linear combination of levels of Factor B.

The sum of squares for $A(B)$ is

$$SS_{A(B)} = n\sum_{j}\sum_{k}(M_{jk} - M_k)^2$$
$$= 25[(3.32)^2 + (8.66)^2 + (-3.32)^2 + (-8.66)^2] = 4300.90$$
$$= SS_A + SS_{AB}.$$

At each of the K levels of Factor B there are $(J - 1)$ degrees of freedom for variation between levels of Factor A. The number of degrees of freedom for $A(B)$ is therefore $K(J - 1)$. The $A(B)$ effect accounts for the $(J - 1)$ degrees of freedom for the A main effect as well as the $(J - 1)(K - 1)$ degrees of freedom for interaction.

The $[A(B) + B]$ summary table for the Sleep deprivation × NVH data set is shown in Table 4.6.

Error rates and critical constants

An ANOVA-model analysis of a two-factor design assigns a conventional error rate (α) to each of the three families of contrasts (A, B and AB) defined on the effect parameters of the model, thereby allowing the overall per-experiment error rate (PEER, the expected number of errors in the analysis as a whole) to reach 3α. As Betz and Levin (1982) pointed out, the sum of the error rates assigned to standard families in a standard analysis (3α) can be partitioned in a nonstandard analysis in the same way that between-cell variation is partitioned. The $A(B)$ family includes two orthogonal sub-families (A and AB), each of which is assigned an error rate of α in a standard ANOVA-model analysis. In order to ensure that an $[A(B) + B]$ analysis is not inherently more conservative than a standard analysis, Betz and Levin suggested that the family-based error rate for $A(B)$ contrasts should be set at 2α. If $\alpha = .05$, the Betz and Levin CC for Bonferroni-t CIs on the $k = 4$ factorial contrasts in the $A(B)$ family ($\psi_{A(b_1)}$, $\psi_{A(b_2)}$, ψ_A and ψ_{AB}) is

$$CC_{A(B)} = t_{2\alpha/(2k_{A(B)});96} = t_{.10/8;96} = 2.277.$$

Table 4.6 *ANOVA summary table for [A(B) + B] analysis*

Source	*SS*	*df*	*MS*	*F*	*p*
A(B) [Sleep Dep (NVH)]	4300.90	2	2150.45	93.25	< .001
B (NVH)	272.25	1	272.25	11.81	< .001
Error	2213.76	96	23.06		
Total	6786.91	99			

The Bonferroni-*t* procedure controls the *A*(*B*) per-family error rate (PFER, the expected number of noncoverage errors in a family of *k* CIs) exactly at 2α, the sum of the error rates for the CIs on ψ_A and ψ_{AB} in an ANOVA-model analysis.

The CC for the single contrast in the standard *B* main effect 'family' is $CC_B = \sqrt{F_{.05;\,1,96}} = t_{.025;\,96} = 1.985$.

Note that the slightly larger CC for *A*(*B*) contrasts will produce wider CIs for the *A* main effect and *AB* interaction contrasts in the analysis based on the *A* simple effects model than for the same contrasts in the ANOVA-model analysis, where the CC is 1.985 for all intervals. This reduction in precision is the trade-off for access to inference on *A* simple effect contrasts, which is not possible in the standard analysis.

The FWER for A(B) contrasts As is always the case, the Bonferroni-*t* procedure controls the PFER exactly and the FWER conservatively. The degree of conservatism depends on the extent of the redundancy (if any) within the family of contrasts. If the two simple effect contrasts were to be excluded from the planned set (leaving only the orthogonal contrasts ψ_A and ψ_{AB}), then the Bonferroni-*t* CC would be $t_{2\alpha/(2k_{A(B)});96} = t_{.10/4;96} = t_{.025;\,96}$, the CC used in the ANOVA-model analysis. In that analysis, the *A*(*B*) FWER for inferences on those two contrasts would be close to $1-(1-\alpha)^2 = .0975$, the FWER for two statistically independent inferences. Adding the two simple effect contrasts introduces linear dependence into the family of contrasts (because the added contrasts can be expressed as linear combinations of the two contrasts already in the family), and reduces further the FWER from Bonferroni-*t* SCIs.

The Scheffé procedure The CC for Scheffé CIs on *A*(*B*) contrasts is

$$CC_{A(B)} = \sqrt{K(J-1)F_{\phi;\,K(J-1),N-JK}}, \tag{4.7}$$

where ϕ is the FWER chosen for the *A*(*B*) family. Betz and Levin (1982) suggest that the FWER (a probability) should be partitioned using the principle described above for the Bonferroni-*t* procedure, so that the FWER would be set at $2\alpha = .10$. The rationale for this approach is problematic, however, because it treats the expected number of errors in the family as though it is a probability, and in some applications the distinction between the two can be far from trivial. If the Scheffé procedure is to be used for the purpose of controlling the FWER for the *A*(*B*) family of contrasts in a two-factor design, then ϕ [in (4.7)] should be set at a value no greater than $1-(1-\alpha)^2 = .0975$, the upper limit of the FWER for this particular family. This is close to the value of $2\alpha = .10$ recommended by Betz and Levin, but in some other applications the maximum FWER is much lower than the PFER.

With $\phi = .0975$, the Scheffé CC is

$$CC_{A(B)} = \sqrt{2F_{.0975;2,96}} = 2.184,$$

which is slightly smaller than the Bonferroni-*t* CC of 2.277, so the Scheffé procedure should be preferred to the Bonferroni-*t* procedure in this case.

Contrast statistics and CIs from the [*A*(*B*) + *B*] analysis are given in Table 4.7. With the exception of the *B* main effect contrast, all of the CIs have different limits from those in the ANOVA-model analysis. The ψ_A and ψ_{AB} intervals are 10% wider in the [*A*(*B*) + *B*] analysis.

The *A* simple effects model would be ideal if the experimenter regarded NVH as a factor of secondary interest. If NVH had been included as a factor in the design only to see whether it influences the size of the sleep deprivation effect, then it would be perfectly reasonable for the experimenter to interpret the interaction as a difference in sleep deprivation simple effects rather than as a difference in NVH simple effects, and to have no interest in *B* simple or main effects. In this case the *A*(*B*) family would include all of the contrasts of interest.

What if all factorial effects are equally important?

Unlike *A* and *B* main effects, *A* and *B* simple effects are not mutually orthogonal. The correlations between simple effect contrasts in a balanced 2 × 2 design are as follows:[2]

	$\psi_{A(b_1)}$	$\psi_{A(b_2)}$	$\psi_{B(a_1)}$	$\psi_{B(a_2)}$
$\psi_{A(b_1)}$	1.0			
$\psi_{A(b_2)}$	0	1.0		
$\psi_{B(a_1)}$	0.5	–0.5	1.0	
$\psi_{B(a_2)}$	–0.5	0.5	0	1.0

In order to carry out a coherent analysis including inferences on *A* and *B* simple

Table 4.7 *Confidence intervals from [A(B)+B] analysis*

Contrast	$\hat{\psi}_g$	$\hat{\sigma}_{\hat{\psi}_g}$	Confidence interval
$\psi_{A(b_1)}$	6.640	1.358	$\psi_{A(b_1)} \in (3.673,\ 9.607)$
$\psi_{A(b_2)}$	17.320	1.358	$\psi_{A(b_2)} \in (14.353,\ 20.287)$
ψ_A	11.980	0.960	$\psi_A \in (9.882,\ 14.078)$
ψ_{AB}	–10.680	1.921	$\psi_{AB} \in (-15.572, -5.788)$
ψ_B	3.300	0.960	$\psi_B \in (1.394,\ 5.206)$

effect contrasts, we must use a model with a single set of parameters such as the cell means model (4.1), rather than a model with more than one independent set of effect parameters such as the two-factor ANOVA model (4.3) or a simple effects model (4.6). We can, however, carry out a factorial analysis by restricting our attention to the seven factorial contrasts (four simple effect contrasts, two main effect contrasts and one interaction effect contrast) defined in Table 4.1. This is not a trivial restriction, because it excludes 18 of the 25 $\{m, r\}$ contrasts that can be defined on four means, such as $\mu_{11} - (\mu_{12} + \mu_{22})/2$. Most nonfactorial contrasts would make very little sense. Some nonfactorial contrasts can be interpreted [$(\mu_{11} - \mu_{22})$ is the combined effect of sleep deprivation and NVH], but not as an effect of either of the two factors.[3]

A single family containing all factorial contrasts covers all of the between-cell variation that would otherwise be covered by an ANOVA-model analysis or an $[A(B) + B]$ analysis, each of which allows for a PEER of 3α. In order to ensure that an analysis including all factorial contrasts in a single family is not inherently more conservative than either of these analyses, the PFER for the single-family analysis can be set at 3α. The Bonferroni-t CC (with $k = 7$) is

$$CC_{All} = t_{3\alpha/(2\times7);96} = t_{.15/14;96} = 2.339.$$

The Scheffé CC is

$$CC_{All} = \sqrt{(JK-1)F_{\phi;\,JK-1,\,N-JK}} = \sqrt{3F_{\phi;\,3,96}}$$

where $\phi = 1 - (1 - \alpha)^3$ is the maximum *experimentwise* error rate (EWER: the probability of one or more noncoverage errors in the analysis as a whole) from a two-factor ANOVA-model analysis. When $\alpha = .05$, $\phi = .1426$ and the Scheffé CC is

$$CC_{All} = \sqrt{3F_{.1426;\,3,96}} = 2.358.$$

The Scheffé CC is only 1% larger than the Bonferroni-t CC, so the two procedures would produce very similar CIs.

Example 4.1 Constructing CIs on all factorial contrasts

PSY can carry out the nonstandard analysis we are currently examining relatively easily. The group-membership variable in the data section of the *PSY* input file *SD×NVH.in* has values of 1, 2, 3 or 4 referring respectively to cells a_1b_1, a_1b_2, a_2b_1 and a_2b_2. The contrasts section of the file is as follows:

```
[BetweenContrasts]
1  0 -1  0 A(b1)
0  1  0 -1 A(b2)
1 -1  0  0 B(a1)
```

```
0  0  1 -1 B(a2)
1  1 -1 -1 A
1 -1  1 -1 B
1 -1 -1  1 AB
```

When the Analysis Options screen appears, select *Bonferroni t* and set the *Confidence Level* at 100(1 – 3α) = 85%. With these settings and the default scaling option (Mean Difference Contrasts) in place, *PSY* will produce appropriate scaled CIs for all simple and main effect contrasts, but not for the interaction contrast. Output from this first analysis provides the following information about coefficient scaling:

```
              Bonferroni 85% Simultaneous Confidence Intervals
              ------------------------------------------------
              The CIs refer to mean difference contrasts,
               with coefficients rescaled if necessary.
              Rescaled Between contrast coefficients
        Contrast          Group...
                            1           2           3           4
A(b1)         B1          1.000       0.000      -1.000       0.000
A(b2)         B2          0.000       1.000       0.000      -1.000
B(a1)         B3          1.000      -1.000       0.000       0.000
B(a2)         B4          0.000       0.000       1.000      -1.000
A             B5          0.500       0.500      -0.500      -0.500
B             B6          0.500      -0.500       0.500      -0.500
AB            B7          0.500      -0.500      -0.500       0.500
```

With the exception of the *AB* interaction contrast, the scaling of all contrasts agrees with that shown in Table 4.1. The mean difference default scaling option in *PSY* is appropriate for simple and main effect contrasts from factorial designs of any degree of complexity. This scaling is always inappropriate for any interaction contrast to be interpreted as a mean difference between levels of one factor in the magnitude of a mean difference between levels of another factor. Having run the first analysis, hit the green Run Analysis button to bring up the Analysis Options screen again, then select the *Interaction Contrasts* scaling option. The *Between order* option will be highlighted, and you will now be in a position to specify the *order* of the interaction contrast(s) you wish to rescale. Any interaction contrast in a two-factor design is a *first-order* interaction contrast, so it is necessary to set the order equal to 1 to achieve the appropriate scaling for ψ_{AB}. (The default order of zero produces mean difference scaling.)

The output file will now include two analyses differing only in the scaling of contrast coefficients. Output from the second analysis includes the following:

```
      Bonferroni 85% Simultaneous Confidence Intervals
      ------------------------------------------------
The coefficients are rescaled if necessary
to provide a metric appropriate for interaction contrasts.
Order of interaction for Between contrasts: 1
```

Rescaled Between contrast coefficients

	Contrast	Group... 1	2	3	4
A(b1)	B1	2.000	0.000	-2.000	0.000
A(b2)	B2	0.000	2.000	0.000	-2.000
B(a1)	B3	2.000	-2.000	0.000	0.000
B(a2)	B4	0.000	0.000	2.000	-2.000
A	B5	1.000	1.000	-1.000	-1.000
B	B6	1.000	-1.000	1.000	-1.000
AB	B7	1.000	-1.000	-1.000	1.000

As a consequence of the adoption of a scaling appropriate for first-order interaction contrasts, $\sum c_{jk}^{+} = 2.0$ for all contrasts in this analysis. This scaling agrees with that given in Table 4.1 for the *AB* interaction contrast, but for no other contrast.

To produce a CI table with the correct scaling for all contrasts, edit the output file by replacing the *AB* row from the first analysis with the corresponding row from the second analysis. The edited table of raw CIs is as follows:

Raw CIs (scaled in Dependent Variable units)

	Contrast	Value	SE	..CI limits.. Lower	Upper
A(b1)	B1	6.640	1.358	3.464	9.816
A(b2)	B2	17.320	1.358	14.144	20.496
B(a1)	B3	-2.040	1.358	-5.216	1.136
B(a2)	B4	8.640	1.358	5.464	11.816
A	B5	11.980	0.960	9.734	14.226
B	B6	3.300	0.960	1.054	5.546
AB	B7	-10.680	1.921	-15.172	-6.188

The CC for this analysis (2.339) is 18% larger than the CC for the analysis based on the two-factor ANOVA model (1.985). The decrease in the precision of estimates of the main effect and interaction contrasts is the price paid for the inclusion of the four simple effect contrasts in the analysis.

Note that because this single-family analysis uses the same CC for all CIs in the analysis, any differences in interval width are due entirely to differences between contrasts in their standard errors. This (perhaps desirable) feature of the single-family analysis is not shared with other approaches to the analysis of factorial designs. As we will see in Chapter 5, multifactor ANOVA-model analyses often use different CCs for different families of contrasts, thereby introducing variation across families in precision of estimation.

Increasing the complexity of factorial designs

In this chapter we have restricted our attention to the two-factor factorial design with only two levels on both factors. As we have seen, there is more than one way of carrying out a coherent analysis of data from this most simple of factorial designs, because the most appropriate analysis depends on the importance attached by the experimenter to inferences about simple effects parameters, parameters not defined by the two-factor ANOVA model.

When a factor in a two-factor design has more than two levels, it becomes possible to define more than one contrast across factor levels. All of the SCI procedures discussed in Chapter 2 can be used to evaluate contrasts referring to main effect parameters, and some can be used to evaluate contrasts referring to interaction parameters. The discussion in Chapter 5 of analyses involving multilevel factors assumes familiarity with the material in Chapter 2 as well as the material in this chapter.

It is possible, of course, to design factorial experiments with more than two factors. We consider multifactor experiments briefly in Chapter 5.

Further reading

Most analyses of factorial designs reported in the research literature use sequential procedures that treat tests on simple effects as 'follow-up' tests if (and only if) the hypothesis of no interaction is rejected. Levin and Marascuilo (1972), Marascuilo and Levin (1976) and Rosnow and Rosenthal (1989) discuss the problems associated with this approach. See Betz and Gabriel (1978) for a detailed discussion of models allowing for coherent analyses including simple effects as well as interaction effects.

Questions and exercises

1. The text file *Ch4 Q1.txt* contains data from a 2×2 experiment with $n = 40$ observations per group. (The first variable in the file has values of 1, 2, 3 or 4 referring respectively to cells a_1b_1, a_1b_2, a_2b_1 and a_2b_2.) The experimenter decides to analyse the data by carrying out planned tests of null hypotheses on the following three orthogonal contrasts: $A(b_1)$, $A(b_2)$ and B. The tests are carried out with a per-contrast Type I error rate of .05. The ANOVA summary table for this analysis is as follows:

Source	*SS*	*df*	*MS*	*F*	*p*
$A(b_1)$	80.000	1	80.000	9.286	.003
$A(b_2)$	20.000	1	20.000	2.321	.130
B	490.000	1	490.000	56.875	< .001
Error	1344.000	156	8.615		

Because $A(b_1)$ (with a raw sample value of 2.0) is statistically significant whereas $A(b_2)$ (with a sample value of 1.0) is not, the experimenter concludes (among other things) that the *A* effect is greater at b_1 than at b_2.

(a) Is the experimenter's conclusion justified by the analysis? If not, why not?

(b) Use *PSY* to construct planned individual 95% CIs on the same set of three contrasts. What does this analysis suggest about
(i) the size of the *A* effect at b_1 and at b_2
(ii) the difference between these two simple effects?

(c) Carry out an alternative analysis allowing for inferences on all of the contrasts of interest to the experimenter, including the difference between the two *A* simple effects. Justify your choice of error rate.

Notes

1. Although it might appear that there are $(J + K)$ zero-sum constraints on interaction parameters, there are only $(J + K - 1)$ linearly independent constraints. The last constraint ($\sum_k \alpha\beta_{jk} = 0$ for the last level of Factor *A*) can be deduced from the earlier constraints.

2. In an equal-*n* experiment, the correlation between two contrasts (which is also the correlation between the two coefficient vectors) is the correlation between the contrast sample values across an indefinitely large number of replications of the experiment.

3. To say that the combined effect of two factors is large is to say nothing about the nature or magnitude of an effect of either factor. A combined effect of ϕ could arise from an *A* main effect of ϕ together with no *B* effects, a *B* main effect of ϕ together with no *A* effects, a nonzero *A* main effect combining additively with a nonzero *B* main effect, or various interactive (non-additive) *A* and *B* combinations. Similarly, a zero combined effect could result from a range of patterns of individual effects, including compensatory additive main effects (such as an *A* main effect of ϕ and a *B* main effect of $-\phi$).

5 Complex Factorial Designs

A factorial analysis of data from a $J \times K$ design can be based on any of the models (cell means, main effects, saturated two-factor ANOVA, or simple effects) discussed in Chapter 4. Multiplicity issues, similar to those discussed in the context of single-factor designs in Chapter 2, arise when factors have multiple levels. If $J > 2$ and $K > 2$ (that is, if both factors have more than two levels), then each of the effects defined by the two-factor ANOVA model (the *A* and *B* main effects and the *AB* interaction effect) has more than 1 degree of freedom, and it is therefore possible to define multiple contrasts within each of the three families of contrasts. Modified versions of all of the CI procedures discussed in Chapter 2 can be used to construct SCIs on main effect parameters of the two-factor ANOVA model. Most (but not all) of these procedures can also be used to construct SCIs on interaction parameters.

When both factors have multiple levels, the interaction means account for at least half of the degrees of freedom between cells in an ANOVA-model analysis. In a 3×4 design, for example, 6 of the 11 degrees of freedom between cells are accounted for by the interaction means. It follows that a contrasts analysis based on the saturated two-factor ANOVA model is likely to be much more complex than an analysis based on the main effects (no interaction) model. The simpler model is therefore a very attractive option if the degree of heterogeneity in interaction parameters (as quantified by f_{AB}) is trivially small.

Partitioning variation, degrees of freedom and the overall error rate

Table 5.1 shows the effects defined by four saturated models of data from a balanced $J \times K$ design. The analysis based on each model accounts for all of the $(JK - 1)$ degrees of freedom for variation between cells.

An analysis based on the two-factor ANOVA model partitions between-cell variation into three orthogonal multiple-*df* components (two main effects and one interaction), thereby defining three orthogonal families of contrasts. In a standard factorial analysis an error rate of α is assigned to the inference (or set of inferences) about each effect, so the overall PEER (the expected number of errors in the analysis) is 3α.

Table 5.1 *Effects and sub-effects in four analyses of data from a $J \times K$ design*

Model	Effects	Sub-effects	*df*	Expected number of errors
Two-factor	A		$J-1$	α
	B		$K-1$	α
	AB		$(J-1)(K-1)$	α
$A(B)+B$	$A(B)$		$K(J-1)$	2α
		$A(b_k)$		
		A		
		AB		
	B		$K-1$	α
$B(A)+A$	$B(A)$		$J(K-1)$	2α
		$B(a_j)$		
		B		
		AB		
	A		$J-1$	α
Cell means	*All*		$JK-1$	3α
		$A(b_k)$		
		$B(a_j)$		
		A		
		B		
		AB		

The $A(B)$ effect in an $[A(B) + B]$ analysis includes A simple effects as well as the A main effect and the AB interaction effect, as sub-effects. Following Betz and Levin (1982), we will assign a PFER of 2α (the sum of the error rates traditionally assigned to A and AB) to the $A(B)$ effect. Similarly, a $[B(A) + A]$ analysis combines the B and AB components of the standard partition, thereby defining an effect $[B(A)]$ that includes B simple effects as well as the B main effect and the AB interaction effect.

The final model shown in Table 5.1 is the cell means model, which does not lead to an analysis based on multiple families of contrasts. This should be the preferred model for experimenters who wish to leave open the possibility of basing their analysis on inferences on any set of factorial effects, including A and B simple effects.

The error-rate conventions proposed by Betz and Levin (1982) for nonstandard analyses are about as reasonable as the established conventions for standard analyses, provided that they refer to expected numbers of errors rather than probability-based error rates. In practice, this means that the figures in the final column of Table 5.1 can be used as nominal error rates in planned analyses using the Bonferroni-t procedure, but slightly smaller nominal error rates should be used for the Scheffé procedure and any other procedure that controls the FWER directly rather than indirectly via control over the PFER.

For Scheffé analyses, the FWER should be set at $1 - (1 - \alpha)^2$ when the PFER for a Bonferroni-t analysis would have been set at 2α, and at $1 - (1 - \alpha)^3$ when the PFER would have been set at 3α.

The Fee × Treatment data set

Consider a randomized experiment designed to determine whether the magnitude of the fee paid for treatment has any influence on the effectiveness of four treatments intended to help smokers reduce cigarette consumption. Factors and factor levels are

A (Fee)	a_1:	\$100 fee
	a_2:	\$50 fee
	a_3:	no fee
B (Treatment)	b_1:	a treatment emphasizing social supports
	b_2:	a treatment emphasizing lifestyle changes
	b_3:	hypnosis
	b_4:	education concerning smoking and health.

Treatments b_1 and b_2 (only) are considered to be behavioural treatments. The dependent variable is the difference between the number of cigarettes smoked per day before treatment and the corresponding number 12 months after the completion of treatment.

The data set analysed below was produced in a simulation of this experiment, with $n = 16$ subjects per cell ($N = 192$ subjects in all). The data are contained in the text file *fee×treatment.txt*, so you can carry out alternative analyses and compare the outcomes with those discussed here. You can also compare the estimated parameters with the actual parameter values used in the simulation (population means and error variance) given in the *Questions and exercises* section at the end of this chapter. Sample means and standard deviations are given in Table 5.2.

A two-factor ANOVA-model analysis First, we consider an analysis based on the two-factor ANOVA model. This analysis is presented first because it is a CI-analysis version of the analysis strategy that is almost universally used to analyse data from $J \times K$ factorial experiments. The two-way ANOVA summary table is shown in Table 5.3(a). This table shows that each of the three sources of between-cell variation has more than 1 degree of freedom, so it will be

Table 5.2 *Means (and standard deviations) from the Fee ×Treatment data set*

	b_1 Social	b_2 Lifestyle	b_3 Hypnosis	b_4 Education	M_j
a_1 \$100	23.88 (4.19)	25.44 (8.98)	25.25 (8.20)	18.25 (6.80)	23.20
a_2 \$50	25.44 (10.35)	21.44 (8.59)	23.69 (6.18)	19.50 (6.57)	22.52
a_3 zero	26.00 (7.59)	23.75 (7.56)	2.63 (6.63)	12.06 (7.60)	16.11
M_k	25.10	23.54	17.19	16.60	

necessary to define multiple contrasts within each effect in order to carry out an exhaustive contrasts analysis. Given the F statistic for interaction (F_{AB} = $F_{6,180}$ = 11.08), *STATISTICA Power Analysis* produces a 90% CI on f_{AB} of (0.432, 0.697). We can infer from this CI that there are substantial differences among the interaction effect parameters, so the main effects model is untenable. We can expect interaction contrasts to play an important role in the interpretation of the data.

Inferences on contrasts in an analysis consistent with 90% two-sided CIs on one or more of the three f parameters (f_A, f_B and f_{AB}), or with .05-level tests of the three homogeneity hypotheses (H_A, H_B and H_{AB}) must be based on Scheffé SCIs or tests. We will return to the treatment of factorial contrasts after we examine the summary tables from alternative analyses.

Alternative analyses The remaining summary tables in Table 5.3 show the partitions (if any) of between-cell variation in analyses allowing for inferences on simple effects. The [*A*(*B*) + *B*] analysis allows for inferences on Fee simple effects (but not for inferences on Treatment simple effects), as well as the Fee main effect and the Fee × Treatment interaction. The [*B*(*A*) + *A*] analysis allows for inferences on Treatment simple effects at any level of the Fee factor (but not for inferences on Fee simple effects), the Treatment main effect and the Fee × Treatment interaction.

The final analysis in Table 5.3 allows for direct inferences on all factorial effects, including Fee simple effects and Treatment simple effects. In this analysis there is no partition of between-cells variation, so the single F ratio is identical to that obtained from a single-factor ANOVA.

Table 5.3 *Summary tables from four analyses of the Fee × Treatment data set*

(a) *Two-way ANOVA summary table*

Source	*SS*	*df*	*MS*	*F*	*p*
A (Fee)	1959.12	2	979.56	17.02	< .001
B (Treatment)	2714.52	3	904.84	15.73	< .001
A × *B*	3825.38	6	637.56	11.08	< .001
Error	10,356.69	180	57.54		
Total	18,855.71	191			

(b) *Summary table for [A(B)+B] analysis*

Source	*SS*	*df*	*MS*	*F*	*p*
A(*B*)	5784.50	8	723.06	12.57	< .001
B	2714.52	3	904.84	15.73	< .001
Error	10,356.69	180	57.54		
Total	18,855.71	191			

(c) *Summary table for [B(A)+A] analysis*

Source	*SS*	*df*	*MS*	*F*	*p*
B(*A*)	6539.90	9	726.66	12.63	< .001
A	1959.12	2	979.56	17.02	< .001
Error	10,356.69	180	57.54		
Total	18,855.71	191			

(d) *Summary table for one-way analysis*

Source	*SS*	*df*	*MS*	*F*	*p*
Between cells	8499.02	11	979.56	17.02	< .001
Error	10,356.69	180	57.54		
Total	18,855.71	191			

The first and fourth of the summary tables in Table 5.3 can be obtained from any statistical package that supports ANOVA. The *SS*s and *df*s for the remaining tables can be obtained by adding relevant figures from the two-way ANOVA summary table. [$SS_{A(B)} = SS_A + SS_{AB}$; $SS_{B(A)} = SS_B + SS_{AB}$. Degrees of freedom for *A*(*B*) and *B*(*A*) are obtained in the same way.]

Factorial contrasts for complex two-factor designs

Most of the CI procedures appropriate for contrasts analyses in single-factor designs can also be applied to contrasts within the families defined by whatever model is chosen for the analysis of a two-factor design. There are, however, some additional issues to be considered, arising from the fact that most (usually all) of the factorial contrasts of interest to experimenters can be expressed as *product contrasts*. A product contrast has a contrast coefficient vector referring to cell means that can be expressed as a product of a coefficient vector referring to levels of Factor A and a coefficient vector referring to levels of Factor B. It is important for two reasons to understand the implications of the fact that factorial contrasts can usually be expressed as product contrasts. First, a set of product contrasts can usually provide a comprehensive and comprehensible account of effects that might otherwise be regarded as impenetrable, such as an interaction effect when both factors have more than two levels. Second, a recently developed SCI procedure designed for post hoc analyses of sets of product contrasts often provides greater precision (and never provides poorer precision) than the Scheffé procedure.

Product contrasts

Suppose that the experimenter decides to base the analysis of the data from the Fee × Treatment experiment on contrasts with the following coefficient vectors referring to factor levels:

	Levels of Factor A				Levels of Factor B			
	a_1	a_2	a_3		b_1	b_2	b_3	b_4
$\mathbf{c}'_{A_1}$	[0.5	0.5	−1.0]	$\mathbf{c}'_{B_1}$	[0.5	0.5	−0.5	−0.5]
$\mathbf{c}'_{A_2}$	[1.0	−1.0	0]	$\mathbf{c}'_{B_2}$	[1.0	−1.0	0	0]
				$\mathbf{c}'_{B_3}$	[0	0	1.0	−1.0]

Two of these coefficient vectors define complex contrasts on factor levels: $\mathbf{c}'_{A_1}$ defines a {2, 1} contrast on levels of Factor A concerned with the average effect of paying a fee for treatment; and $\mathbf{c}'_{B_1}$ defines a {2, 2} contrast on levels of Factor B concerned with the average difference between behavioural and non-behavioural treatments. The remaining coefficient vectors define {1, 1} contrasts (comparisons) on factor levels.

Two points should be noted about these coefficient vectors. First, they could well have been different. The experimenters have the same kind of choice about how to define contrasts on levels of multilevel factors (including choices about the number of contrasts and relationships between them) as do experimenters defining contrasts on group means in the analysis of a single-factor design. This is not true of a two-level factor, where the only possible contrast on factor levels is a comparison. Second, the definitions of the contrasts are incomplete. A contrast coefficient vector referring to levels of Factor A might be part of the definition of an A main effect contrast, an A simple effect contrast, an AB interaction contrast, or even an A contrast on a linear combination of levels of B such as $(b_1+b_2)/2$. One way to complete the definition is to specify an additional coefficient vector referring to levels of B, together with a rule showing how the A and B coefficient vectors are to be combined to produce a set of contrast coefficients referring to the 12 cells of the experimental design. The rule is as follows:

- the A coefficient vector (written as a column rather than a row) is placed to the left of a $J \times K$ matrix (a 3×4 matrix in this case);
- the B coefficient vector is placed above the matrix;
- each entry in the $J \times K$ *product coefficient matrix* is obtained by multiplying the row coefficient from the A vector by the column coefficient from the B vector.

This procedure is illustrated below, where a product coefficient matrix $\mathbf{C}_{A_1B_2}$ is defined as a product of the coefficient vectors $\mathbf{c}_{A_1}$ and $\mathbf{c}_{B_2}$:

$$\begin{array}{ccc} & \mathbf{c}'_{B_2} & \\ & \Downarrow & \\ & \begin{bmatrix} 1.0 & -1.0 & 0 & 0 \end{bmatrix} & \\ \mathbf{c}_{A_1} \Rightarrow \begin{bmatrix} 0.5 \\ 0.5 \\ -1.0 \end{bmatrix} & \begin{bmatrix} 0.5 & -0.5 & 0 & 0 \\ 0.5 & -0.5 & 0 & 0 \\ -1.0 & 1.0 & 0 & 0 \end{bmatrix} & \Leftarrow \mathbf{C}_{A_1B_2} \;\; (c_{A_1B_{2_{jk}}} = c_{A_{1_j}} \times c_{B_{2_k}}). \end{array}$$

In the terminology of matrix algebra, $\mathbf{C}_{A_1B_2} = \mathbf{c}_{A_1}\mathbf{c}'_{B_2}$ is the *matrix product* of the A and B coefficient vectors. The contrast coefficients in the matrix $\mathbf{C}_{A_1B_2}$ refer to the cell means in the matrix

$$\boldsymbol{\mu} = \begin{bmatrix} \mu_{11} & \mu_{12} & \mu_{13} & \mu_{14} \\ \mu_{21} & \mu_{22} & \mu_{23} & \mu_{24} \\ \mu_{31} & \mu_{32} & \mu_{34} & \mu_{44} \end{bmatrix}.$$

The resulting contrast is

$$\begin{aligned} \psi_{A_1B_2} &= \sum_j \sum_k c_{A_1B_{2_{jk}}} \mu_{jk} \\ &= 0.5\mu_{11} - 0.5\mu_{12} + 0.5\mu_{21} - 0.5\mu_{22} - \mu_{31} + \mu_{32} \\ &= 0.5(\mu_{11} - \mu_{12}) + 0.5(\mu_{21} - \mu_{22}) - (\mu_{31} - \mu_{32}) \\ &= \left(\psi_{B_2(a_1)} + \psi_{B_2(a_2)} \right)/2 - \psi_{B_2(a_3)}. \end{aligned}$$

The same contrast can be written as

$$\psi_{A_1B_2} = (0.5\mu_{11} + 0.5\mu_{21} - \mu_{31}) - (0.5\mu_{12} + 0.5\mu_{22} - \mu_{32})$$
$$= \psi_{A_1(b_1)} - \psi_{A_1(b_2)}.$$

This is the interaction contrast concerned with the average effect of paying a fee (A_1) on the size of the difference between the outcomes of the social supports treatment and the lifestyle treatment (B_2). As is always the case with an *AB product interaction contrast*, $\psi_{A_1B_2}$ is concerned with an *A* difference in a *B* difference, or equivalently, a *B* difference in an *A* difference.

It is possible to summarize these operations in matrix algebra terminology without referring directly to $\mathbf{C}_{A_1B_2}$. The contrast $\psi_{A_1B_2}$ can be defined as

$$\psi_{A_1B_2} = \mathbf{c}'_{A_1} \boldsymbol{\mu}\, \mathbf{c}_{B_2}. \qquad (5.1)$$

The form of this expression implies that this particular contrast is a *double* linear combination of cell means whose definition depends on two coefficient vectors, the left coefficient vector referring to levels of Factor *A*, the right coefficient vector referring to levels of Factor *B*. [If you are not familiar with matrix multiplication, be assured that (5.1) is simply a compact way of repeating what has already been stated about the definition of the product interaction contrast $\psi_{A_1B_2}$.]

When at least one of the factors in a two-factor ANOVA model has only two levels (that is, in $2 \times K$ and $J \times 2$ designs), any interaction contrast can be expressed as a product contrast. When both factors have more than two levels, however, it is possible to define interaction contrasts that are not product contrasts. For example, the coefficient matrix

$$\mathbf{C} = \begin{bmatrix} 2 & -1 & -1 & 0 \\ -1 & 2 & -1 & 0 \\ -1 & -1 & 2 & 0 \end{bmatrix}$$

defines a genuine *AB* interaction contrast because the coefficients sum to zero within each row and column, but this matrix cannot be expressed as the product of two vectors. In general, interaction contrasts that cannot be expressed as product contrasts are not readily interpretable (they certainly cannot be interpreted as *A* differences in *B* differences) and are usually of little or no interest to experimenters.[1] We will therefore restrict our attention to factorial contrasts that are also product contrasts. Given this restriction, we can always interpret all factorial contrasts of interest in terms of relatively simple coefficient vectors referring to factor levels, rather than in terms of relatively complex coefficient matrices referring to cell means.

Main effect contrasts as product contrasts All main effect contrasts can be expressed as product contrasts. Any A main effect contrast has a coefficient matrix of the form

$$\mathbf{C}_{A_g B_0} = \mathbf{c}_{A_g} \mathbf{c}'_{B_0} \tag{5.2}$$

where $\mathbf{c}_{A_g}$ is a contrast coefficient vector referring to levels of Factor A,

and $\mathbf{c}'_{B_0}$ is a vector with K identical elements.

If the coefficient vectors are scaled so that

$$\sum_j c^+_{A_h} = \sum_k c^+_{B_0} = 1.0$$

(that is, if the sum of positive values in each coefficient vector is 1.0), then the resulting product coefficient matrix defines a A main effect mean difference contrast. For example, if

$$\mathbf{c}'_{B_0} = [\ 0.25 \ \ 0.25 \ \ 0.25 \ \ 0.25\ ] \qquad (\sum_k c^+_{B_0} = 1.0),$$

then the elements of the coefficient matrix $\mathbf{C}_{A_1 B_0} = \mathbf{c}_{A_1}\mathbf{c}'_{B_0}$ are obtained by multiplication as follows:

$$\begin{array}{c}
\mathbf{c}'_{B_0} \\
\Downarrow
\end{array}$$

$$\begin{array}{cc}
 & [\ 0.25 \quad 0.25 \quad 0.25 \quad 0.25\ ] \\
\mathbf{c}_{A_1} \Rightarrow \begin{bmatrix} 0.5 \\ 0.5 \\ -1.0 \end{bmatrix} & \begin{bmatrix} 0.125 & 0.125 & 0.125 & 0.125 \\ 0.125 & 0.125 & 0.125 & 0.125 \\ -0.250 & -0.250 & -0.250 & -0.250 \end{bmatrix} \Leftarrow \mathbf{C}_{A_1 B_0}.
\end{array}$$

The mean difference contrast defined by the coefficient matrix $\mathbf{C}_{A_1 B_0}$ is

$$\begin{aligned}
\psi_{A_1 B_0} &= \mathbf{c}'_{A_1} \boldsymbol{\mu}\, \mathbf{c}_{B_0} \\
&= \frac{\mu_{11}+\mu_{12}+\mu_{13}+\mu_{14}+\mu_{21}+\mu_{22}+\mu_{23}+\mu_{24}}{8} - \frac{\mu_{31}+\mu_{32}+\mu_{33}+\mu_{34}}{4} \\
&= \frac{\mu_{1.}+\mu_{2.}}{2} - \mu_{3.}.
\end{aligned}$$

As is always the case with product contrasts, one way of interpreting this contrast is via interpretations of the coefficient vectors (referring to factor levels) whose product defines the coefficient matrix (referring to cell means). We have already seen that $\mathbf{c}_{A_1}$ is concerned with the effect of charging a fee for treatment. The coefficient vector $\mathbf{c}'_{B_0}$ referring to levels of Factor B is not a contrast coefficient vector, because the coefficients do not sum to zero ($\sum c_{B_0} = 1.0$). Each coefficient is $1/K$, so the function of $\mathbf{c}'_{B_0}$ is to average across the K levels of Factor B. Thus the contrast $\mathbf{c}'_{A_1}\boldsymbol{\mu}\,\mathbf{c}_{B_0}$ is the *average* effect of charging a fee, averaging across treatments.

If $\psi_{A_1B_0}$ is an A main effect contrast, it must be possible to apply the coefficients in $\mathbf{c}_{A_1}$ to the A main effect parameters defined by the two-factor ANOVA model. Recall that $\boldsymbol{\alpha}' = [\alpha_1\ \alpha_2\ \cdots\ \alpha_J]$ is the vector of A main effect parameters. The contrast $\psi_{A_1B_0}$ can be written as

$$\psi_{A_1B_0} = \mathbf{c}'_{A_1}\boldsymbol{\alpha} = \frac{\alpha_1 + \alpha_2}{2} - \alpha_3. \tag{5.3}$$

Thus we can define an A main effect contrast from a 3×4 design as a linear combination of the three A main effect parameters defined by the two-factor ANOVA model, or as a linear combination of the 12 cell means defined by the cell means model. The first of these alternatives offers advantages associated with the simplicity of an expression like (5.3). The second alternative also offers advantages: as we have seen, it is easier to define simple effect contrasts on parameters of the cell means model than it is on parameters of the two-factor ANOVA model.

B main effect contrasts can be defined in a similar manner. The coefficient vector $\mathbf{c}'_{A_0} = [0.\dot{3}\ \ 0.\dot{3}\ \ 0.\dot{3}]$ has J coefficients, each of which is $1/J$, so the function of $\mathbf{c}'_{A_0}$ in an expression like

$$\psi_{A_0B_1} = \mathbf{c}'_{A_0}\boldsymbol{\mu}\,\mathbf{c}_{B_1}$$

is to ensure that $\psi_{A_0B_1}$ averages across levels of Factor A. It follows that if $\mathbf{c}_{B_1}$ is a contrast coefficient vector (that is, if $\sum c_{B_1} = 0$), then $\psi_{A_0B_2}$ is a B main effect contrast.

Simple effect contrasts as product contrasts Any A simple effect contrast can be expressed as a product contrast by combining an A contrast coefficient vector with a B vector with a single nonzero coefficient of 1.0, such as

$$\mathbf{c}'_{b_1} = [\ 1.0\quad 0\quad 0\quad 0\ ].$$

The function of $\mathbf{c}_{b_1}$ is to ensure that a product contrast like $\mathbf{c}'_{A_1}\boldsymbol{\mu}\,\mathbf{c}_{b_1}$ is an A simple effect contrast at the first level of Factor B, as can be seen from the implied coefficient matrix referring to cell means:

$$\begin{array}{ccc} & \mathbf{c}'_{b_1} & \\ & \Downarrow & \\ & [\ 1.0\quad 0\quad 0\quad 0\ ] & \\ \mathbf{c}_{A_1} \Rightarrow \begin{bmatrix} 0.5 \\ 0.5 \\ -1.0 \end{bmatrix} & \begin{bmatrix} 0.5 & 0 & 0 & 0 \\ 0.5 & 0 & 0 & 0 \\ -1.0 & 0 & 0 & 0 \end{bmatrix} & \Leftarrow \mathbf{C}_{A_1b_1}. \end{array}$$

The contrast

$$\psi_{A_1(b_1)} = \mathbf{c}'_{A_1}\,\boldsymbol{\mu}\,\mathbf{c}_{b_1}$$
$$= \frac{\mu_{11} + \mu_{21}}{2} - \mu_{31}$$

is the effect of charging a fee on the outcome of the treatment emphasizing social supports.

Any A simple effect contrast can be expressed as a double linear combination of the cell means where

- the left (A) coefficient vector is a contrast coefficient vector, and
- the right (B) coefficient vector has a single nonzero coefficient of 1.0.

Any B simple effect contrast can be expressed as a double linear combination of the cell means where

- the left (A) coefficient vector has a single nonzero coefficient of 1.0, and
- the right (B) coefficient vector is a contrast coefficient vector.

Subset effect contrasts The 2 × 2 design is a simple design not only because there can be only one contrast for each family defined by the two-factor ANOVA model (that is, one A main effect, one B main effect and one AB interaction contrast), but also because traditional factorial contrasts (main effect, interaction effect and simple effect contrasts) are the only product contrasts likely to be of interest to experimenters. This is not necessarily true of complex two-factor designs.

Consider the family of A product contrasts where

- $\mathbf{c}_A$ is a vector of contrast coefficients, and
- $\mathbf{c}_B$ is a vector containing m positive coefficients $(1 \le m \le K)$, each with a value of $1/m$, and $(K - m)$ zero coefficients.

Every contrast in this family is a difference between levels of Factor A, averaged across some (at least one, possibly all) of the levels of Factor B. All A simple effect contrasts (m = 1) and A main effect contrasts (m = K) belong to this family. If m has an intermediate value $(1 < m < K)$, then the contrast is an A *subset effect* contrast concerned with a difference between levels of A, averaged across a subset of levels of B.

If the experimenters were interested in estimating the value of the interaction contrast $\psi_{A_1B_1}$ (the effect of paying a fee on the average difference in outcome between behavioural and nonbehavioural treatments), then they would probably also be interested in estimating the effect of paying a fee on the average outcome of behavioural treatments, a contrast averaging across two of the four levels of Factor B. The B coefficient vector required to express this as a product contrast is $\mathbf{c}'_{B_4} = [0.5 \;\; 0.5 \;\; 0 \;\; 0]$, so the product coefficient matrix referring to cell means is obtained as follows:

$$\begin{array}{cc} & \mathbf{c}'_{B_4} \\ & \Downarrow \\ & [\;0.5 \quad 0.5 \quad 0 \quad 0\;] \\ \mathbf{c}_{A_1} \Rightarrow \begin{bmatrix} 0.5 \\ 0.5 \\ -1.0 \end{bmatrix} & \begin{bmatrix} 0.25 & 0.25 & 0 & 0 \\ 0.25 & 0.25 & 0 & 0 \\ -0.50 & -0.50 & 0 & 0 \end{bmatrix} \Leftarrow \mathbf{C}_{A_1B_4}, \end{array}$$

defining the mean difference contrast

$$\begin{aligned} \psi_{A_1B_4} &= \mathbf{c}'_{A_1} \boldsymbol{\mu} \, \mathbf{c}_{B_4} \\ &= \frac{\mu_{11} + \mu_{12} + \mu_{21} + \mu_{22}}{4} - \frac{\mu_{31} + \mu_{32}}{2}. \end{aligned}$$

Although this is not a simple, main or interaction effect contrast, it is a factorial contrast in the sense that one coefficient vector defines a difference between levels of one factor, while the other coefficient vector specifies the levels of the other factor across which that difference is averaged.

It is possible, of course, to define *B* product contrasts that are not simple, main or interaction effect contrasts. If the experimenters were interested in the difference between behavioural and nonbehavioural treatment outcomes for fee-paying participants, they could define the *A* coefficient vector $\mathbf{c}'_{A_3} = [0.5 \;\; 0.5 \;\; 0]$, then estimate the value of the *B* subset effect contrast $\mathbf{c}'_{A_3} \boldsymbol{\mu} \, \mathbf{c}_{B_1}$.

Families of product contrasts Each of the effects defined by one of the models in Table 5.1 is the basis for a family of factorial contrasts. All product contrasts belonging to all of these families are double linear combinations of the form $\mathbf{c}'_A \boldsymbol{\mu} \, \mathbf{c}_B$ where at least one of the two coefficient vectors contains contrast coefficients. Product contrasts belonging to the families defined by all but one of the models (those leading to analyses that partition variation between cells) must satisfy additional restrictions:

Family of product contrasts	*Additional restrictions on coefficient vectors*	
	$\mathbf{c}_A$	$\mathbf{c}_B$
A	$\Sigma c_A = 0$	All c_B +ve, equal
B	All c_A +ve, equal	$\Sigma c_B = 0$
AB	$\Sigma c_A = 0$	$\Sigma c_B = 0$
A(*B*)	$\Sigma c_A = 0$	None
B(*A*)	None	$\Sigma c_B = 0$

All product contrasts are members of the final family ('*All*') defined in Table 5.1. Some of these families of product contrasts include others as sub-families. For example, any product contrast with $\sum c_A = 0$ is a member of the $A(B)$ family. It follows that all A and AB contrasts are also members of the $A(B)$ family. The final family listed above (*All*) includes all of the other families as sub-families. The sub-families (if any) included within each family are shown in Table 5.1 under the heading *sub-effects*.

Simplifying the terminology for factorial contrasts

The terminology we have been using to this point to refer to factorial contrasts

$$\psi_{A_g B_h} = \mathbf{c}'_{A_g} \boldsymbol{\mu}\, \mathbf{c}_{B_h}$$

makes it clear that all factorial contrasts of interest are likely to be product contrasts. We will now adopt a more simple and conventional terminology to refer to factorial contrasts. Names for the most common types of factorial contrasts are as follows:

Contrast type	*Contrast name*	*Coefficient vectors*	
		A	B
A main effect	A_g	$\mathbf{c}_{A_g}$	$\mathbf{c}_{B_0}$
B main effect	B_h	$\mathbf{c}_{A_0}$	$\mathbf{c}_{B_h}$
AB interaction	$A_g B_h$	$\mathbf{c}_{A_g}$	$\mathbf{c}_{B_h}$
A simple effect	$A_g(b_k)$	$\mathbf{c}_{A_g}$	$\mathbf{c}_{b_k}$
B simple effect	$B_h(a_j)$	$\mathbf{c}_{a_j}$	$\mathbf{c}_{B_h}$

Note that the name given to a main effect contrast makes no reference to the 'ignored' factor (with a coefficient vector containing uniform coefficients); the contrast that might have been called $A_1 B_0$ is simply called A_1.

Critical constants for CIs on contrasts within families

The CCs for Bonferroni-*t* CIs within families are

$$\mathrm{CC}_{fam} = t_{r\alpha/(2k_{fam});\nu_\varepsilon}$$

where r is the number of standard sub-families (A, B, AB) combined to define the family in question ($r = 1$, 2 or 3);

and k_{fam} is the number of planned contrasts in the family in question.

The CCs for simultaneous Scheffé CIs within families are

$$CC_{fam} = \sqrt{\nu_{fam} F_{\phi;\nu_{fam},\nu_\varepsilon}}$$

where ν_{fam} is the number of degrees of freedom for the family in question

and $\phi = 1-(1-\alpha)^r$.

Values of ν_{fam} can be obtained from the *df* column in Table 5.1.

In analyses based on multiple families (including the standard ANOVA-model analysis based on three families), each family has its own CC_{fam}. The size of the CC usually varies across families.

Selecting factorial contrasts on a post hoc basis

If contrasts within families are to be selected on a post hoc basis, it is generally useful to examine a *profile plot* (sometimes called an interaction plot) showing the profile of *b* means at each level of *A* (such as that shown in Figure 5.1) or a plot showing the profile of *a* means at each level of *B*, before deciding on which contrasts to use as the basis for the analysis.

Figure 5.1 (based on the means in Table 5.2) shows the profile of treatment means at each level of the Fee factor. If the profiles were parallel, there would be no evidence of interaction. The two nonzero fee levels produce similar profiles, with no suggestion of substantial effects of the magnitude of the fee. The zero-fee profile has a very different shape, however, with a very low mean for the hypnosis treatment and a relatively low mean for the education treatment.

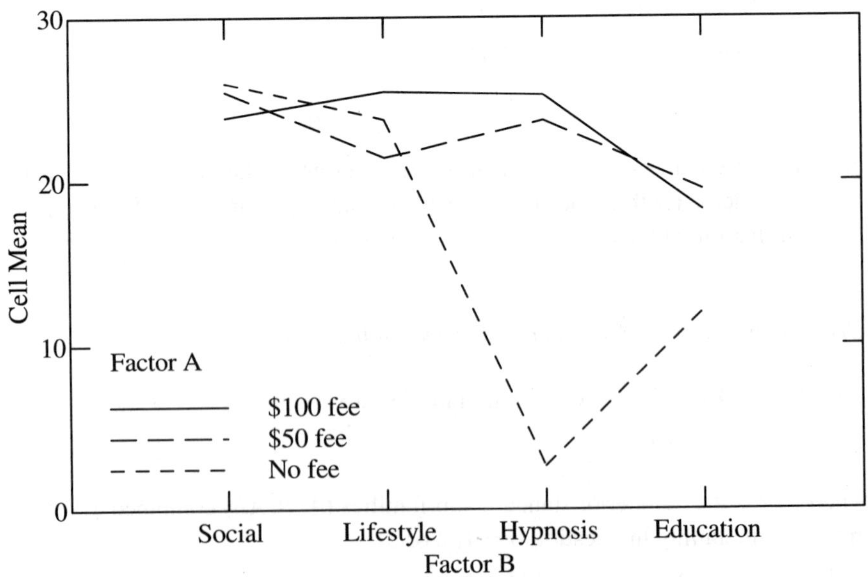

Figure 5.1 *Profiles of treatment means from the Fee × Treatment data set*

It is clear from the row and column means in Table 5.2 that a zero vs nonzero fee contrast (A_1) would account for most of the variation in the A main effect, and a behavioural (social and lifestyle) vs nonbehavioural (hypnosis and education) contrast (B_1) would account for most of the variation in the B main effect. Furthermore, the pattern of the departures from parallelism in Figure 5.1 suggests that the product interaction contrast A_1B_1 should account for a substantial proportion of the variation between interaction means.

The remaining degrees of freedom can be accounted for without redundancy by defining main effect contrasts orthogonal to those already defined. A_2, comparing the two nonzero fee levels, is the only A main effect contrast orthogonal to A_1, which averages across those two levels. There is, however, more than one way of defining two B main effect contrasts orthogonal to B_1 and to each other. Whereas an inspection of the column means in Table 5.2 suggests that the difference between hypnosis and education (B_3) is unlikely to contribute noticeably to variation in the B main effect, Figure 5.1 suggests that the simple effect for this comparison when no fee is charged is very different from the average of the simple effects for the same comparison when a fee is charged. We would therefore expect the interaction contrast A_1B_3 to account for some of the variation between interaction means. B_2 is the only B main effect contrast orthogonal to both B_1 and B_3. We now have a basis for an exhaustive analysis of variation between cell means, in which both of the contrast coefficient vectors used in the A main effects analysis will be combined with each of the three contrast coefficient vectors used in the B main effects analysis to define six product interaction contrasts, thereby accounting for all of the interaction variation. In a balanced design, orthogonality of main effect contrasts implies orthogonality of all of the associated interaction contrasts. We therefore expect to be able to account for most of the nontrivial variation between cell means in terms of a small subset of an exhaustive set of orthogonal main effect and interaction contrasts.

Example 5.1 An F-based two-way ANOVA

We now complete the F-based analysis that began with the construction of a 90% two-sided CI on f_{AB} (page 94), the lower limit of which provided evidence of substantial differences between interaction parameters, thereby confirming the need to base the contrasts analysis on the saturated two-factor ANOVA model (as distinct from the unsaturated main effects model).

PSY recognizes only one between-subjects factor, because it is based on a means model rather than a factorial ANOVA model. It is therefore necessary to provide the program with contrast coefficients referring to cell means rather than factor levels.

The contrasts section of the file *feextreatment.in* contains coefficients referring to the 12 cell means in the order $\mu_{11}, \mu_{12}, \cdots, \mu_{34}$. That is, the second subscript (referring to levels of Factor B) changes faster than the first subscript (referring to levels of Factor A). Given this convention, the first four coefficients $(c_{11}, c_{12}, c_{13}, c_{14})$ are those in the first row of the relevant $\mathbf{C}$ matrix, the next four $(c_{21}, c_{22}, c_{23}, c_{24})$ are those in the second row of that matrix, and so on. The same convention is used for the group-membership variable in the data section of the file, so the values of that variable vary from 1 to 12, where 1 refers to cell $(a_1 b_1)$, 2 refers to cell $(a_1 b_2)$, and so on. The contrasts are defined as follows:

```
[BetweenContrasts]
 1  1  1  1    1  1  1  1   -2 -2 -2 -2 A1
 1  1  1  1   -1 -1 -1 -1    0  0  0  0 A2
 1  1 -1 -1    1  1 -1 -1    1  1 -1 -1 B1
 1 -1  0  0    1 -1  0  0    1 -1  0  0 B2
 0  0  1 -1    0  0  1 -1    0  0  1 -1 B3
 1  1 -1 -1    1  1 -1 -1   -2 -2  2  2 A1B1
 1 -1  0  0    1 -1  0  0   -2  2  0  0 A1B2
 0  0  1 -1    0  0  1 -1    0  0 -2  2 A1B3
 1  1 -1 -1   -1 -1  1  1    0  0  0  0 A2B1
 1 -1  0  0   -1  1  0  0    0  0  0  0 A2B2
 0  0  1 -1    0  0 -1  1    0  0  0  0 A2B3
```

The additional spaces in each row (which have no effect on the program) are boundaries between levels of Factor A. Each A main effect coefficient (referring to a level of A) appears $K = 4$ times (once for each level of B), and each B main effect coefficient (referring to a level of B) appears $J = 3$ times. Coefficients for the interaction contrasts are obtained by multiplication.

A main effect analysis To carry out an F-based A main effect analysis, it is necessary to cut the irrelevant B and AB contrasts from the input file, select *User-supplied Critical Constants* from the Analysis Options menu, and provide the required Scheffé CC ($\text{CC}_A = \sqrt{2F_{.05;2,180}} = 2.4683$). The raw SCIs are as follows:

```
Special Confidence Intervals: User-supplied Critical Constants
   Between main effect CC: 2.4683
   -------------------------------------------------------------

   Raw CIs (scaled in Dependent Variable units)
   -------------------------------------------------------------
   Contrast          Value          SE            ..CI limits..
                                                 Lower        Upper
   -------------------------------------------------------------
   A1                6.750          1.161         3.884        9.616
   A2                0.688          1.341        -2.622        3.997
   -------------------------------------------------------------
```

B main effect analysis After all but the B main effect contrasts have been cut from the input file, the program is provided with the Scheffé CC for B main effect contrasts ($\text{CC}_B = \sqrt{3F_{.05;3,180}} = 2.8221$). The resulting SCIs are

```
Special Confidence Intervals: User-supplied Critical Constants
   Between main effect CC: 2.8221
   ---------------------------------------------------------
   Raw CIs (scaled in Dependent Variable units)
   ---------------------------------------------------------
   Contrast         Value          SE          ..CI limits..
                                              Lower       Upper
   ---------------------------------------------------------
   B1               7.427        1.095        4.337      10.517
   B2               1.563        1.548       -2.807       5.932
   B3               0.583        1.548       -3.786       4.953
   ---------------------------------------------------------
```

AB interaction analysis The default scaling option (inappropriate for the six interaction contrasts) must be changed by selecting *Interaction Contrasts* from the Analysis Options menu and setting the (highlighted) *Between order* to 1. The user-supplied CC is $CC_{AB} = \sqrt{6F_{.05;6,180}} = 3.5910$. The SCIs are

```
Special Confidence Intervals: User-supplied Critical Constants
   Between main effect CC: 3.591
   ---------------------------------------------------------
   The coefficients are rescaled if necessary
    to provide a metric appropriate for interaction contrasts.
   Order of interaction for Between contrasts: 1

   Raw CIs (scaled in Dependent Variable units)
   ---------------------------------------------------------
   Contrast         Value          SE          ..CI limits..
                                              Lower       Upper
   ---------------------------------------------------------
   A1B1           -15.156        2.323      -23.496      -6.816
   A1B2            -1.031        3.285      -12.826      10.764
   A1B3            15.031        3.285        3.236      26.826
   A2B1             1.063        2.682       -8.568      10.693
   A2B2            -5.563        3.793      -19.182       8.057
   A2B3             2.813        3.793      -10.807      16.432
   ---------------------------------------------------------
```

Interpretation Suppose that the experimenters were willing to interpret the raw CIs, because the dependent variable units (reduction in number of cigarettes smoked per day) are meaningful. Suppose further that a difference of about 5 (corresponding to a reduction of 35 cigarettes per week) was interpreted as the smallest clinically significant difference on this scale. Given these guidelines, we can now interpret the Scheffé CIs.

Directional inference is possible for contrasts A_1, B_1, A_1B_1 and A_1B_3. It is clear that payment of a fee increases the average effectiveness of the treatments (A_1), but it is not clear whether the size of this effect is clinically important. The magnitude of the fee effect is greater for nonbehavioural than behavioural treatments (A_1B_1), and it is greater for hypnosis than for education (A_1B_3). Although both of these differences in effect size are estimated with poor precision, the first is clearly clinically significant, and both may be substantial. Finally, averaging across all three levels of Factor A, the behavioural treatments are more effective than the nonbehavioural treatments (B_1). This difference is not large but it is not trivially small. Although the CI on B_3 appears to justify the conclusion that hypnosis and education are practically equivalent, it must be remembered that this main effect contrast is an average of heterogeneous simple effect contrasts, as shown by the CI on A_1B_3.

Comments on the analysis This ANOVA-model analysis provides inferences only on contrasts within main effect and interaction families associated with the parameters of the model. These inferences raise a number of questions about the simple effects of paying a fee [in particular $A_1(b_3)$ and $A_1(b_4)$], and about simple effect comparisons between hypnosis and education at each level of the fee factor. Thus although the analysis is exhaustive (in the sense that it accounts for all of the variation between cell means), it does not provide answers to some questions of interest.

The analysis is coherent in the sense that all inferences within a given family make appropriate use of the same critical value of the relevant F distribution. Consequently, the FWERs for these families cannot exceed the nominated error rate $(\alpha = .05)$ and the overall EWER cannot exceed $1-(1-\alpha)^3 = .143$, provided that the assumptions underlying the ANOVA model are satisfied. Many analyses reported in the literature (such as analyses including 'follow-up' tests on simple effects if H_{AB} is rejected) are not coherent in this sense.

A planned analysis would almost certainly have produced substantial improvements in precision. It does not follow, however, that the resulting analysis would have been more satisfying to the experimenters, because it is unlikely that the contrasts planned for the analysis would have been identical to those chosen after an inspection of the data. Indeed, the A contrast coefficient vectors in a planned analysis would probably have been $\mathbf{c}'_{A_3} = [1 \ \ 0 \ {-1}]$ and $\mathbf{c}'_{A_4} = [0.5 \ {-1.0} \ \ 0.5]$, referring to the linear and quadratic components of trend across fee levels. (Trend analysis is discussed in Appendix D.) Trend contrasts across levels of a quantitative factor (in this case fee in dollars) usually work well if the regression of outcome on factor levels turns out to be approximately linear. This is not the case with the current data set. It is also unlikely that a contrast like A_1B_3, an important component of the observed interaction, would have been included in a planned analysis.

The experimenters could, of course, have included all $\{m, r\}$ main effect contrasts and all associated interaction contrasts in a 'planned' analysis including all such contrasts that might conceivably turn out to be of interest. This strategy does not produce a reduction in CI width, however, because the number of planned contrasts ($k_A = 6$, $k_B = 25$ and $k_{AB} = 150$) is so large that all of the Bonferroni-t CCs are larger than the corresponding Scheffé CCs.

The sensitivity of the analysis of interaction could be improved slightly by replacing the test statistic (F_{AB}), which does not take into account the fact that all interaction contrasts in this analysis are product contrasts, with a test statistic that does take this into account. The *F*-based Scheffé SCI procedure is always unnecessarily conservative in analyses of product interaction contrasts when both factors have more than two levels (Boik, 1993). The studentized maximum root (*SMR*) procedure developed by Boik (1986) provides smaller CCs for product interaction contrasts, although the resulting increase in precision in this case would be modest, as it usually is when $(J - 1) = 2$ or $(K - 1) = 2$.

The *SMR* approach has a much wider role in the analysis of factorial designs than has previously been recognized. As we will see, *SMR* SCIs can be much more precise than Scheffé SCIs in analyses that include inferences on simple effects.

The *SMR* procedure

The *F*-based Scheffé procedure allows for direct inferences on all contrasts in the relevant families, including those that cannot be expressed as product contrasts. As we have seen, whenever both factors have more than two levels some interaction contrasts cannot be expressed as product contrasts. It follows that when $J > 2$ and $K > 2$ the Scheffé procedure is unnecessarily conservative for post hoc analyses of product contrasts within any family that includes interaction contrasts (that is, all families except for the *A* and *B* families). Boik (1986, 1993) showed that an SCI procedure based on the *studentized maximum root* (*SMR*) distribution can provide SCIs that control the FWER for product interaction contrasts in balanced designs, and that this procedure always provides greater precision than the Scheffé CC when both factors have more than two levels. Although the rationale for the *SMR* procedure assumes a balanced design, Boik (1993) has shown with Monte Carlo methods that the procedure also works well with unbalanced designs. Bird and Hadzi-Pavlovic (2003) have shown that the *SMR* procedure can be extended to control FWERs for post hoc analyses of product contrasts within nonstandard families [*A*(*B*), *B*(*A*) and *All*] in balanced designs.

Parameters of the SMR distribution The *SMR* distribution has three degrees-of-freedom parameters: *p*, *q* and *df*. For our purposes, $df = \nu_E$. The *p* parameter for a particular family of product contrasts is a function of the number of levels of Factor *A* and/or the restrictions imposed on the $\mathbf{c}_A$ coefficient vectors contributing to the definition of the contrasts in that family. For families with no restrictions on $\mathbf{c}_A$ vectors [namely *B*(*A*) and *All*], $p = J$ (the number of levels of *A*). For families where $\mathbf{c}_A$ must be a contrast coefficient vector [*A*, *AB* and *A*(*B*)], $p = J - 1$. For the *B* family where $\mathbf{c}_A$ must be a vector of identical positive values, $p = 1$. Similarly, the *q* parameter for a particular family of product contrasts is a function of the number of levels of Factor *B* and/or the restrictions imposed on $\mathbf{c}_B$ coefficient vectors. For families with no restrictions on $\mathbf{c}_B$ vectors [*A*(*B*) and *All*], $q = K$ (the number of levels of *B*). For families where $\mathbf{c}_B$ must be a contrast coefficient vector [*B*, *AB* and *B*(*A*)], $q = K - 1$. For the *A* family where $\mathbf{c}_B$ must be a vector of identical positive values, $q = 1$.

With one exception (the family of all product contrasts), the product of the *SMR* p and q parameters for a particular family is the degrees-of-freedom parameter ν_1 for the *F* distribution used by the Scheffé procedure. Values of p, q and ν_1 are shown in Table 5.4. The *SMR* CC for $100(1-\phi)\%$ SCIs on product contrasts in a particular family is

$$\mathrm{CC}_{SMR} = \sqrt{SMR_{\phi;\,p,q,\nu_E}}\,. \tag{5.4}$$

Critical values of the *SMR* distribution for standard α levels have been tabled by Boik (1986, 1993) and Harris (1994). These tables do not, however, provide the critical values required for nonstandard FWERs. Any required *SMR* critical value can be obtained from the *PSY* Probability Calculator.

SMR or Scheffé? The relationship between Scheffé (*F*) and *SMR* CCs can be illustrated by calculating both for the various families that might be considered as a basis for analysing data from the Fee × Treatment study. The CCs for $\alpha = .05$ are shown in Table 5.5. The final column shows the ratio of the two CCs, which is also the ratio of CI half-widths, with the *SMR* CC in the numerator. With the exception of the two main effect families, the *SMR* procedure provides smaller CCs (and therefore greater precision) than the *F*-based Scheffé procedure. It is worth noting that the overall sample size required for the Scheffé procedure to provide the same precision for an analysis including all product contrasts as the *SMR* procedure with $N = 192$ is approximately

Table 5.4 *Degrees-of-freedom parameters of SMR and F distributions required for the construction of simultaneous confidence intervals within families*

Family	*SMR parameters*		*F parameter*
	p	q	ν_1
A	$J-1$	1	$J-1$
B	1	$K-1$	$K-1$
AB	$J-1$	$K-1$	$(J-1)(K-1)$
A(*B*)	$J-1$	K	$K(J-1)$
B(*A*)	J	$K-1$	$J(K-1)$
All	J	K	$JK-1$

Table 5.5 *Critical constants for Scheffé (F) and SMR simultaneous confidence intervals within families for 3 × 4 design with $\nu_2 = 180$*

Family	*Critical constants*		$\dfrac{CC_{SMR}}{CC_F}$
	Scheffé (*F*)	*SMR*	
A	$\sqrt{2F_{.05;2,180}} = 2.468$	$\sqrt{SMR_{.05;\,2,1,180}} = 2.468$	1.000
B	$\sqrt{3F_{.05;3,180}} = 2.822$	$\sqrt{SMR_{.05;\,1,3,180}} = 2.822$	1.000
AB	$\sqrt{6F_{.05;6,180}} = 3.591$	$\sqrt{SMR_{.05;\,2,3,180}} = 3.315$	0.923
A(*B*)	$\sqrt{8F_{.0975;8,180}} = 3.704$	$\sqrt{SMR_{.0975;\,2,4,180}} = 3.332$	0.900
B(*A*)	$\sqrt{9F_{.0975;9,180}} = 3.884$	$\sqrt{SMR_{.0975;\,3,3,180}} = 3.399$	0.875
All	$\sqrt{11F_{.1426;11,180}} = 4.034$	$\sqrt{SMR_{.1426;3,4,180}} = 3.525$	0.874

$$192\left(\frac{CC_{F_{All}}}{CC_{SMR_{All}}}\right)^2 = 251.5,$$

an increase of 31%.

Whenever $p = 1$ or $q = 1$, all contrasts in the family are necessarily product contrasts and the *SMR* CC is identical to the Scheffé CC. (For this reason, *SMR* tables and the *PSY* Probability Calculator begin with critical values of $SMR_{2,2,\nu}$.) When both $p \geq 2$ and $q \geq 2$, the maximal product contrast in the family invariably accounts for less of the variation between cells than the maximal contrast in the same family, so the *F*-based Scheffé procedure (which controls the FWER for inferences on all contrasts) is unnecessarily conservative for an analysis restricted to product contrasts. When applied to contrasts within any family except the family of all product contrasts, the *SMR* procedure is optimal for analyses where the only restriction is that all contrasts within the family must be product contrasts. This restriction poses no real problems in practice, because factorial contrasts of interest to experimenters are almost invariably product contrasts.

When applied to the family of all product contrasts, the *SMR* distribution with parameters $p = J$ and $q = K$ produces a CC that is slightly larger than the (unknown) optimal CC for post hoc product contrasts. (As pointed out earlier, this is the only case where $pq \neq \nu_1$.) The problem here is that while these values of p and q allow for the possibility that either of the coefficient vectors ($\mathbf{c}_A$ and $\mathbf{c}_B$) may be unrestricted, they do not take into account the fact that one must contain contrast coefficients that sum to zero. As a consequence, this particular application of the *SMR* procedure is conservative, producing wider CIs than are necessary to control the FWER at the nominated level.

The 2×2 design is the only factorial design where the *SMR* procedure has no legitimate role. In a 2×2 design the *SMR* procedure for the family of all product contrasts (with $p = 2$ and $q = 2$) is so conservative [because $(pq = 4) > (\nu_1 = 3)$] that it is inferior to the Scheffé procedure. For every other family of factorial contrasts in a 2×2 design [including *A*(*B*) and *B*(*A*)], the Scheffé and *SMR* CCs are identical.

For every other two-factor design, the *SMR* CC for all product contrasts is smaller (sometimes only marginally smaller, as is the case with the 2×3 design) than the Scheffé CC. With the sole exception of the 2×2 design, the *SMR* procedure should be preferred to the *F*-based Scheffé procedure for inferences on product contrasts in any family where both p and q are greater than 1.

A directional inference from an *SMR*-based CI implies that the relevant homogeneity hypothesis can be rejected by the associated *SMR* test (Boik, 1986). A discussion of the relatively complex *SMR* test statistic, a statistic not calculated by statistical packages, is beyond the scope of this book. The *SMR* statistic is not required for the construction of *SMR*-based SCIs, however, just as the ANOVA *F* statistic is not required for the construction of *F*-based Scheffé SCIs. In both cases, all that is required is the relevant CC, which is not a statistic. When the Scheffé and *SMR* CCs differ, there is no justification for the use of *F*-based CIs on *f* parameters or *F* tests of homogeneity of effect parameters in conjunction with *SMR* SCIs. If the *F* test rejects (or does not reject) the relevant homogeneity hypothesis, it does not follow that it is possible (or not possible) to construct at least one *SMR* CI that excludes the value zero.

Example 5.2 SMR SCIs on all factorial contrasts

The following analysis uses the *SMR* procedure to evaluate any factorial contrast of interest to the experimenter. All of the main and interaction effect contrasts from the previous ANOVA-model analysis are included, as well as all of the associated *A* and *B* simple effect contrasts and a *B* subset effect contrast. The additional contrasts in the *PSY* input file are defined as follows:

```
1  0  0  0   1  0  0  0  -2  0  0  0 A1(b1)
0  1  0  0   0  1  0  0   0 -2  0  0 A1(b2)
0  0  1  0   0  0  1  0   0  0 -2  0 A1(b3)
0  0  0  1   0  0  0  1   0  0  0 -2 A1(b4)
1  0  0  0  -1  0  0  0   0  0  0  0 A2(b1)
0  1  0  0   0 -1  0  0   0  0  0  0 A2(b2)
0  0  1  0   0  0 -1  0   0  0  0  0 A2(b3)
0  0  0  1   0  0  0 -1   0  0  0  0 A2(b4)
1  1 -1 -1   0  0  0  0   0  0  0  0 B1(a1)
0  0  0  0   1  1 -1 -1   0  0  0  0 B1(a2)
```

```
0  0  0  0   0  0  0  0   1  1 -1 -1 B1(a3)
1 -1  0  0   0  0  0  0   0  0  0  0 B2(a1)
0  0  0  0   1 -1  0  0   0  0  0  0 B2(a2)
0  0  0  0   0  0  0  0   1 -1  0  0 B2(a3)
0  0  1 -1   0  0  0  0   0  0  0  0 B3(a1)
0  0  0  0   0  0  1 -1   0  0  0  0 B3(a2)
0  0  0  0   0  0  0  0   0  0  1 -1 B3(a3)
0  0  1 -1   0  0  1 -1   0  0  0  0 B3(a1,a2)
```

To carry out an *SMR* analysis, we select the *Maximum root* (*factorial*) option with $p = 3$ and $q = 4$ (from the relevant *SMR* parameter values in Table 5.4) and change the confidence level to $100(1-.05)^3 = 85.74$. These settings have the effect of instructing the program to use a CC of $\sqrt{SMR_{.1426;3,4,180}} = 3.525$ for all contrasts (see Table 5.5). The resulting CIs will be scaled appropriately for all contrasts except the six interaction contrasts. The interaction contrasts are analysed after the default scaling option is changed by selecting *Interaction Contrasts* and setting the (now highlighted) *Between order* to 1.

After appropriate editing of the output that brings the correctly scaled contrasts together in a single table, the raw CIs appear as follows:

```
Maximum root 85.74% Simultaneous Confidence Intervals
  p = 3 and q = 4
---------------------------------------------------------
Raw CIs (scaled in Dependent Variable units)
---------------------------------------------------------
Contrast          Value         SE          ..CI limits..
                                          Lower       Upper
---------------------------------------------------------
A1                6.750      1.161        2.657      10.843
A2                0.688      1.341       -4.039       5.414
A1(b1)           -1.344      2.323       -9.530       6.842
A1(b2)           -0.313      2.323       -8.499       7.874
A1(b3)           21.844      2.323       13.658      30.030
A1(b4)            6.813      2.323       -1.374      14.999
A2(b1)           -1.563      2.682      -11.015       7.890
A2(b2)            4.000      2.682       -5.453      13.453
A2(b3)            1.563      2.682       -7.890      11.015
A2(b4)           -1.250      2.682      -10.703       8.203
B1                7.427      1.095        3.568      11.286
B2                1.563      1.548       -3.895       7.020
B3                0.583      1.548       -4.874       6.041
B1(a1)            2.906      1.896       -3.778       9.590
B1(a2)            1.844      1.896       -4.840       8.528
B1(a3)           17.531      1.896       10.847      24.215
B2(a1)           -1.563      2.682      -11.015       7.890
B2(a2)            4.000      2.682       -5.453      13.453
B2(a3)            2.250      2.682       -7.203      11.703
B3(a1)            7.000      2.682       -2.453      16.453
B3(a2)            4.188      2.682       -5.265      13.640
B3(a3)           -9.438      2.682      -18.890       0.015
B3(a1,a2)         5.594      1.896       -1.090      12.278
A1B2             -1.031      3.285      -12.608      10.546
A2B1              1.063      2.682       -8.390      10.515
A2B2             -5.563      3.793      -18.930       7.805
A2B3              2.813      3.793      -10.555      16.180
---------------------------------------------------------
```

There is a great deal of redundancy in this analysis (with 29 contrasts on 12 means). All of the contrasts used for the ANOVA-model analysis, as well as all simple effect contrasts and the subset effect contrast $B_3\big((a_1+a_2)/2\big)$, are included here in order to allow for detailed comparisons between the two analyses.

The following comments refer to the CIs on simple and subset effect contrasts that could not be included in the ANOVA-model analysis. The CI on $A_1(b_3)$ shows that payment of a fee has a substantial beneficial effect on the outcome of the hypnosis treatment. It is not clear from the CI on $A_1(b_4)$ whether payment of a fee has a beneficial effect on the outcome of the education treatment. The CI on $B_3(a_3)$ shows that when no fee is charged, either the education treatment is superior to hypnosis (possibly by a large margin) or the two treatments are practically equivalent.

While the CI on A_1B_3 implies that the signed difference between the hypnosis and education treatment outcomes (as distinct from the absolute value of the difference) is greater when a fee is charged than when treatment is free, none of the B_3 simple effect CIs implies that these two treatments have different effects at any fee level.[2] Note, however, that whereas the interaction contrast A_1B_3 is the difference between hypnosis and education in the size of the *average* fee effect (averaging across the effects of the \$100 fee and the \$50 fee), the simple effect contrasts $B_3(a_1)$ and $B_3(a_2)$ consider the two nonzero fee levels separately. A more relevant contrast is the average of these two simple effect contrasts

$$B_3\big((a_1+a_2)/2\big) = \frac{B_3(a_1)+B_3(a_2)}{2}$$

for which the CI inference is

$$B_3\big((a_1+a_2)/2\big) \in (-1.09,\ 12.28).$$

The CI on $B_3\big((a_1+a_2)/2\big)$ is tighter than those on $B_3(a_1)$ and $B_3(a_2)$. As a consequence, the subset effect CI provides better evidence than either of the simple effect CIs that education is not a significantly (in the clinical sense) superior treatment to hypnosis when a fee is charged.

Planned contrasts analyses of data from $J \times K$ designs

The Bonferroni-*t* procedure is valid if both the *A* and *B* coefficient vectors for all contrasts to be estimated within a nominated family are defined independently of the data. A Bonferroni-*t* CC depends on the number of planned contrasts in the family in question (k_A, k_B, k_{AB}, $k_{B(A)}$ or k_{All}). In general, $k_{AB} = k_A \times k_B$. If simple effect and/or subset effect contrasts are to be included in the analysis, then the number of planned contrasts in nonstandard families can be large, relative to the number of degrees of freedom for the relevant effect. Nevertheless, Bonferroni-*t* analyses are often more efficient than *F* or *SMR* analyses.

As is the case with a single-factor design, the Bonferroni-*t* procedure should be used only if the reduction it provides in CI width is deemed sufficient to justify the restrictions implied in a planned analysis. We have already seen that Bonferroni-*t* SCIs are actually wider than Scheffé SCIs for analyses of data from a simple 2×2 design. We will now examine the potential advantages (if any) of restricted analyses within families in the case of the 3×4 Fee × Treatment study.

Suppose that the orthogonal *A* contrast coefficient vectors $\mathbf{c}_{A_1}$ and $\mathbf{c}_{A_2}$ and the orthogonal *B* contrast coefficient vectors $\mathbf{c}_{B_1}$, $\mathbf{c}_{B_2}$ and $\mathbf{c}_{B_3}$ were defined before the experiment was run. An ANOVA-model planned analysis based on these coefficient vectors would produce $k_A = 2$ *A* main effect contrasts, $k_B = 3$ *B* main effect contrasts and $k_{AB} = k_A \times k_B = 6$ *AB* interaction contrasts. Because all contrasts within the *A*, *B* and *AB* families are linearly independent, Bonferroni-*t* SCIs for contrasts within these families are necessarily narrower than Scheffé or *SMR* SCIs. The cost of this increase in precision for the contrasts included in the analysis is the range of contrasts excluded from the analysis. Not only are all simple and subset effect contrasts excluded, but so are any other main and interaction effect contrasts that may turn out to be of interest after the data have been inspected.

If the experimenters plan to make direct inferences on simple effect contrasts (as well as main and interaction effect contrasts), then the number of planned contrasts is likely to exceed the number of degrees of freedom between cells. Suppose, for example, that the experimenters plan to add *A* and *B* simple effect contrasts (but not subset effect contrasts) to the set of 11 main effect and interaction contrasts outlined above. There are $Kk_A + Jk_B = 17$ additional contrasts (two *A* simple effect contrasts at each of the four levels of Factor *B* and three *B* simple effect contrasts at each of the three levels of Factor *A*), so the planned analysis must contain considerable redundancy. Nevertheless, the Bonferroni-*t* CC for this planned analysis ($t_{.15/(2\times28);180} = 2.819$) is only 80% as large as the CC required for the unrestricted analysis discussed earlier ($\sqrt{SMR_{.1426;\,3,4,180}} = 3.525$).

If *PSY* is used to construct Bonferroni-*t* intervals in a single-family analysis, *Bonferroni t* is selected and (if appropriate) the confidence level is changed to produce a non-standard PFER such as 2α or 3α. A second run will be required to produce scaling appropriate for interaction contrasts. If the analysis includes more than one family of contrasts, a separate run is needed for each family.

Factorial designs with more than two factors

There is no limit to the number of factors that can be included in a factorial design. Although multifactor ANOVA models can have several sets of effect parameters, they are straightforward extensions of the two-factor ANOVA model. For our purposes it will be sufficient to consider the three-factor design. Suppose that the design of the Sleep deprivation experiment discussed in Chapter 4 is extended to include a third two-level factor, namely Sex (of subject). Factors and factor levels are

Factors	*Factor levels*
A (Sleep deprivation)	a_1: 12 hours of sleep deprivation
	a_2: no sleep deprivation
B (NVH)	b_1: high NVH
	b_2: low NVH
C (Sex of subject)	c_1: male
	c_2: female

Note that unlike the other two factors, Factor C is an *individual difference* or *subject* factor rather than a treatment factor. (Subjects are not randomly assigned to a level of this factor, but they may be randomly sampled from particular populations of males or females.) The three factors are crossed, so the number of cells in the design is $2 \times 2 \times 2 = 8$.

The saturated three-factor ANOVA model for data from a $J \times K \times L$ design is

$$Y_{ijkl} = \mu + \alpha_j + \beta_k + \gamma_l + \alpha\beta_{jk} + \alpha\gamma_{jl} + \beta\gamma_{kl} + \alpha\beta\gamma_{jkl} + \varepsilon_{ijkl} \qquad (5.5)$$

where γ_l is a main effect parameter referring to level l of Factor C ($l = 1, 2, \ldots, L$),

$\alpha\beta_{jk}$, $\alpha\gamma_{jl}$ and $\beta\gamma_{kl}$ are *first-order* (or *double*) interaction parameters,

and $\alpha\beta\gamma_{jkl}$ is a *second-order* (or *triple*) interaction parameter.

Main effect and first-order interaction parameters can be interpreted in much the same way as parameters with similar names in the saturated two-factor ANOVA model (5.2). Note, however, that in the three-factor model

- the α_j parameter refers to an average effect where the averaging takes place across levels of both Factor B and Factor C;
- the $\alpha\beta_{jk}$ parameter refers to an average effect where the averaging takes place across levels of Factor C;
- the $\alpha\beta\gamma_{jkl}$ parameter is the only effect parameter that does not involve any kind of averaging.

When each factor has only two levels, every effect in the model can be interpreted as a contrast on cell means. (This is true of all saturated factorial ANOVA models for 2^q designs: designs with q factors, each of which has two levels.) The cell mean coefficients for a $2 \times 2 \times 2$ design are shown in Table 5.6. The final column of the table shows the sum of the positive coefficients for each contrast. These figures show that the main effect contrasts are scaled as mean difference contrasts and the first-order interaction contrasts are scaled so that each can be interpreted as a difference between two mean difference contrasts. For example, the *AB* contrast is

$$\frac{\mu_{111} + \mu_{112} - \mu_{121} - \mu_{122} - \mu_{211} - \mu_{212} + \mu_{221} + \mu_{222}}{2}$$

$$= \frac{(\mu_{111} + \mu_{112} - \mu_{211} - \mu_{212}) - (\mu_{121} + \mu_{122} - \mu_{221} - \mu_{222})}{2}$$

$$= (\mu_{11.} - \mu_{21.}) - (\mu_{12.} - \mu_{22.}) = A(b_1) - A(b_2).$$

The dot notation used to specify means makes it clear that the *AB* contrast averages across the two levels of Factor *C*.

The cell mean coefficients for the *second-order* interaction contrast *ABC* are scaled so that $\sum c^+_{jkl} = 4.0$. This scaling is required if the contrast is to be interpreted as a difference between the levels of one factor in the magnitude of the *simple interaction* between the other two factors. For example, the *ABC* contrast is

$$(\mu_{111} - \mu_{121} - \mu_{211} + \mu_{221}) - (\mu_{112} - \mu_{122} - \mu_{212} + \mu_{222}) = AB(c_1) - AB(c_2)$$

where *AB* is scaled so that it can be interpreted as a difference between mean difference contrasts. The contrast *ABC* can also be interpreted as the difference between the magnitudes of the simple *AC* interaction contrasts at the two levels

Table 5.6 *Cell mean coefficients for main and interaction effect contrasts in a $2 \times 2 \times 2$ design*

Effect	c_{111}	c_{112}	c_{121}	c_{122}	c_{211}	c_{212}	c_{221}	c_{222}	$\sum c^+$
A	0.25	0.25	0.25	0.25	–0.25	–0.25	–0.25	–0.25	1
B	0.25	0.25	–0.25	–0.25	0.25	0.25	–0.25	–0.25	1
C	0.25	–0.25	0.25	–0.25	0.25	–0.25	0.25	–0.25	1
AB	0.5	0.5	–0.5	–0.5	–0.5	–0.5	0.5	0.5	2
AC	0.5	–0.5	0.5	–0.5	–0.5	0.5	–0.5	0.5	2
BC	0.5	–0.5	–0.5	0.5	0.5	–0.5	–0.5	0.5	2
ABC	1	–1	–1	1	–1	1	1	–1	4

of B, or as the difference between the magnitudes of the simple BC interaction contrasts at the two levels of A. Any second-order interaction contrast from any factorial design must be scaled so that $\sum c^+ = 4.0$ to justify interpretations like these (that is, interpretations implying that the value of the contrast is a difference in a difference between differences).

There are no simple effect parameters in the three-factor ANOVA model. Many types of simple effect parameters can be defined in the context of a cell means model. For example,

- $A(b_k\, c_l)$ is the A simple effect at a particular combination of levels of Factors B and C;
- $A(b_k)$ is the A *simple main effect* at b_k, averaged across levels of Factor C;
- $AB(c_l)$ is the AB *simple interaction effect* at level c_l of Factor C.

Three-factor designs with multiple levels on some factors

Consider now the three-factor design resulting from the addition of a third factor [(C): sex of subject] to the Fee $\times$ Treatment study. The extended $3 \times 4 \times 2$ design has 24 cells. The principles used to calculate the degrees of freedom for individual effects are as follows:

- The number of degrees of freedom for a main effect is one fewer than the number of levels of the relevant factor.
- The number of degrees of freedom for an interaction effect is the product of the degrees of freedom for main effects for all factors involved in the interaction.

The three-way ANOVA partition of the 23 degrees of freedom between cells is shown in Table 5.7.

A three-factor ANOVA-model analysis based on the A and B contrasts discussed earlier would include the following contrasts:

Effect	*Contrasts*
A	A_1, A_2
B	B_1, B_2, B_3
C	C
AB	$A_1B_1, A_1B_2, A_1B_3, A_2B_1, A_2B_2, A_2B_3$
AC	A_1C, A_2C
BC	B_1C, B_2C, B_3C
ABC	$A_1B_1C, A_1B_2C, A_1B_3C, A_2B_1C, A_2B_2C, A_2B_3C$

Table 5.7 *ANOVA partition of between-cells variation in a 3 ×4 ×2 design*

Source	*df*
A (Fee)	2
B (Treatment)	3
C (Sex of subject)	1
AB	6
AC	2
BC	3
ABC	6
Between cells	23

The data section of the *PSY* input file must contain two variables, the first of which is a group-membership variable with values in the range 1 to 24 referring to cells (groups), the second of which is the dependent variable *Y*. The order of the 24 groups should be $a_1b_1c_1$, $a_1b_1c_2$, $a_1b_2c_1$, $\cdots$, $a_3b_4c_2$, with levels of *A* changing most slowly and levels of *C* changing most quickly. The contrasts section of the input file should be similar to the following.

```
[BetweenContrasts]
1  1  1  1  1  1  1  1  1  1  1  1  1  1  1  1 -2 -2 -2 -2 -2 -2 -2 -2 A1
1  1  1  1  1  1  1  1 -1 -1 -1 -1 -1 -1 -1 -1  0  0  0  0  0  0  0  0 A2
1  1  1  1 -1 -1 -1 -1  1  1  1  1 -1 -1 -1 -1  1  1  1  1 -1 -1 -1 -1 B1
1  1 -1 -1  0  0  0  0  1  1 -1 -1  0  0  0  0  1  1 -1 -1  0  0  0  0 B2
0  0  0  0  1  1 -1 -1  0  0  0  0  1  1 -1 -1  0  0  0  0  1  1 -1 -1 B3
1 -1  1 -1  1 -1  1 -1  1 -1  1 -1  1 -1  1 -1  1 -1  1 -1  1 -1  1 -1 C
1  1  1  1 -1 -1 -1 -1  1  1  1  1 -1 -1 -1 -1 -2 -2 -2 -2  2  2  2  2 A1B
.  .
.  .
0  0  0  0  1 -1 -1  1  0  0  0  0 -1  1  1 -1  0  0  0  0  0  0  0  0 A2B3C
```

The entries in the first six (main effect) rows are obtained by repeating the required integer coefficients across levels of irrelevant factors. The entries in all subsequent rows (A1B to A2B3C) are obtained by multiplication.

A *PSY* analysis requires at least three passes through the program, the first to produce CIs on mean difference main effect contrasts, the second to produce intervals on first-order (double) interaction contrasts (A_1B_1 to B_3C) the third to produce intervals on second-order (triple) interaction contrasts (A_1B_1C to A_2B_3C). If SCIs are to be constructed, it is necessary to carry out seven sub-analyses, one for each family of contrasts.

It is easier to construct raw SCIs in between-subjects multifactor ANOVA-model analyses with *SPSS MANOVA* (a program that recognizes distinctions between factors) than with *PSY*. (For details, see Appendix B.) It is usually easier to carry out nonstandard analyses with *PSY*.

Further reading

The treatment of factorial designs in this chapter emphasizes inference from SCIs on contrasts within effects defined by multifactor ANOVA models, or within families defined by alternative models such as simple effect models. Harris (1994, Chapter 4) provides a treatment that is largely compatible with that given here. For the most part Harris discusses directional inference from simultaneous tests, although, as he points out, his approach allows for the construction of SCIs on contrasts. For more traditional treatments of factorial designs with much more emphasis on homogeneity tests, see Kirk (1995), Maxwell and Delaney (1990) or Winer, Brown and Michels (1991). These are large books, containing much more extensive discussions of factorial models than those provided here.

Betz and Levin (1982) discuss coherent analysis strategies for three-factor designs allowing for inferences on simple effects as well as main and interaction effects. Boik (1993) provides a detailed discussion of coherent analyses of first-order interactions, with particular reference to the *SMR* procedure.

Questions and exercises

1. A 2×3 factorial experiment with $n = 20$ observations per cell ($N = 120$ observations in all) includes the following factors and factor levels:

Factors	*Factor levels*
A (Pharmacological treatment)	a_1: Antidepressant medication
	a_2: Placebo
B (Psychological treatment)	b_1: Cognitive behaviour therapy (CBT)
	b_2: Psychodynamic therapy (PT)
	b_3: Minimal-contact control (C).

The analysis is to be based on planned main effect and interaction contrasts. The planned contrast coefficient vectors referring to levels of Factor B are

$$\mathbf{c}'_{B_1} = [0.5 \quad 0.5 \quad -1.0] \qquad \mathbf{c}'_{B_2} = [1.0 \quad -1.0 \quad 0]$$
$$\mathbf{c}'_{B_3} = [1.0 \quad 0 \quad -1.0] \qquad \mathbf{c}'_{B_4} = [0 \quad 1.0 \quad -1.0].$$

(a) Provide an interpretation of the following contrasts:

(i) A
(ii) B_1
(iii) AB_2.

(b) What CCs should be used for 95% SCIs for the A, B and AB families?

(c) Suppose now that the experimenter had decided, before running the study, to include all *B* (but not *A*) simple effect contrasts in the planned analysis, as well as main and interaction effect contrasts.

(i) What families and CCs would be appropriate for this analysis?

(ii) What would you include in the contrasts section of the *PSY* input file?

2. The Fee × Treatment data set (with sample means shown in Table 5.2) was generated in a computer simulation with the following population means:

	b_1 Social	b_2 Lifestyle	b_3 Hypnosis	b_4 Education
a_1 \$100	25	25	25	20
a_2 \$50	25	25	25	20
a_3 zero	25	25	5	10

and a within-populations standard deviation of 8 ($\sigma_\varepsilon^2 = 64$). Given these population means, the population values of the contrasts estimated in the first analysis of the data set (Example 5.1) are as follows:

	B_0	B_1	B_2	B_3
A_0	–	7.5	0	$1.\dot{6}$
A_1	7.5	–15.0	0	10
A_2	0	0	0	0

The population values of standardized contrasts ($\psi_{gh}/\sigma_\varepsilon$) are as follows:

	B_0	B_1	B_2	B_3
A_0	–	0.94	0	0.21
A_1	0.94	–1.88	0	1.25
A_2	0	0	0	0

Contrasts in the first column (with the heading B_0) are *A* main effect contrasts, contrasts in the first row (with the heading A_0) are *B* main effect contrasts and the remaining contrasts are *AB* interaction contrasts.

(a) Of the 11 raw CIs produced in the ANOVA-model Scheffé analysis (Example 5.1), how many produced a noncoverage error?

(b) The approximate standardized SCIs from the same analysis are as follows:

Contrast	Value	SE	..CI limits.. Lower	Upper
A1	0.890	0.153	0.512	1.268
A2	0.091	0.177	-0.346	0.527
B1	0.979	0.144	0.572	1.386
B2	0.206	0.204	-0.370	0.782
B3	0.077	0.204	-0.499	0.653
A1B1	-1.998	0.306	-3.098	-0.899
A1B2	-0.136	0.433	-1.691	1.419
A1B3	1.982	0.433	0.427	3.537
A2B1	0.140	0.354	-1.130	1.410
A2B2	-0.733	0.500	-2.529	1.062
A2B3	0.371	0.500	-1.425	2.166

How many of these standardized intervals produce a noncoverage error?

3. A $2 \times 2 \times 2 \times 2$ design with factors A, B, C and D and $n = 10$ observations per cell is to be analysed with contrasts compatible with a saturated four-factor ANOVA model.

(a) Write down the effects in the model and the number of degrees of freedom for each effect.

(b) What CC would be used for the construction of 95% SCIs controlling the FWER within each of the families defined by the four-factor ANOVA model?

(c) (i) What would you include in the contrasts section of the input file if you were using *PSY* to run the analysis?
(ii) How many runs would be required to carry out the *PSY* analysis? Why?

Notes

1. Exceptions to this generalization can sometimes occur in special cases where the design of the experiment allows for an interpretation of the pattern of the coefficients in the coefficient matrix (Abelson and Prentice, 1997).

2. The CI on A_1B_3 does not imply that the absolute value of the difference between the hypnosis and education treatment outcomes is greater when a fee is charged than when treatment is free, because it is compatible with the possibility that the former difference is small and positive while the latter difference is large and negative.

6 Within-subjects Designs

In a fully randomized experimental design with a single dependent variable, each subject is measured only once. All of the variability in the experiment is derived from differences between different subjects, and the design is therefore referred to as a *between-subjects* design. In a single-factor *repeated measures* design with p conditions, each subject is observed under each condition, so that p measurements (dependent variable scores) are obtained from each subject. If subjects in a treatment outcome study are measured before the treatment (*Pre*), immediately after the treatment (*Post*) and at a follow-up (*FU*) six months after the treatment, then $p = 3$. Variation between the three measurement occasions is based on changes over time within the same subjects, so designs of this kind are often called *within-subjects* designs.

Consider the following set of scores Y_{ij} ($j = 1, 2, \ldots, p$) from a single-factor repeated measures design with $n = 5$ subjects and $p = 3$ measurements per subject:

	Y_{i1} (*Pre*)	Y_{i2} (*Post*)	Y_{i3} (*FU*)
Subject			
1	18	23	19
2	12	15	24
3	12	15	21
4	10	11	18
5	8	11	8
M_j	12.0	15.0	18.0

The traditional approach to the analysis of repeated measures data uses a subjects × measurements univariate ANOVA model (a 5 × 3 model in this case) with one observation per cell.[1] CI analyses based on this model are not recommended in this book. Be aware, however, that most repeated measures analyses reported in the literature are based on the univariate model. The main alternative approach makes use of a *multivariate* analysis of variance (MANOVA) model with less restrictive assumptions.

The multivariate model for single-factor within-subjects designs

The multivariate model recognizes that each subject produces more than one score (three in this case), that those scores may be correlated (subjects with relatively high *Pre* scores may also have relatively high *Post* scores), and that those correlations may vary (the correlation between *Pre* and *Post* may be different from the correlation between *Pre* and *FU,* and both may be different from the correlation between *Pre* and *FU*). The multivariate means model is

$$Y_{ij} = \mu_j + \varepsilon_{ij} \tag{6.1}$$

where Y_{ij} is the score on the dependent variable Y obtained by subject i on measurement j ($j = 1, 2, \ldots, p$),
μ_j is the population mean on measurement j
and ε_{ij} is the error component $(Y_{ij} - \mu_j)$.

It follows from these definitions that the expected value of the error components on each occasion must be zero. The error components are assumed to follow a *multivariate normal* distribution in the population from which subjects are sampled. The most obvious implication of this assumption is that Y scores are normally distributed on each occasion of measurement. A more important implication is that all linear combinations of Y scores are normally distributed.

To illustrate what is (and what is not) implied by the application of the multivariate model to the small data set given above, consider the scores obtained by each subject on the linear combinations $Y_2 - Y_1$ (*Post – Pre*), $Y_3 - Y_2$ (*FU – Post*) and $Y_3 - Y_1$ (*FU – Pre*).

Subject	Y_{i1}	Y_{i2}	Y_{i3}	$Y_{i2} - Y_{i1}$	$Y_{i3} - Y_{i2}$	$Y_{i3} - Y_{i1}$
1	18	23	19	5	–4	1
2	12	15	24	3	9	12
3	12	15	21	3	6	9
4	10	11	18	1	7	8
5	8	11	8	3	–3	0
M	12.0	15.0	18.0	3.0	3.0	6.0
s^2	14.0	24.0	36.5	2.0	36.5	27.5

Each of the linear combinations of the Y variables is itself a variable on which each subject has a score. The linear combination $Y_2 - Y_1$ has a coefficient vector $\mathbf{c}_1' = [-1\ \ 1\ \ 0]$, so the mean score on $Y_2 - Y_1$ is also the sample value of the comparison $\psi_1 = \mu_2 - \mu_1$, namely $\hat{\psi}_1 = M_2 - M_1 = 15.0 - 12.0 = 3.0$. The variance of $Y_2 - Y_1$ scores ($s^2_{Y_2 - Y_1} = 2.0$) is much smaller than the variance of $Y_3 - Y_2$ scores ($s^2_{Y_3 - Y_2} = 36.5$), indicating that changes within individuals between pre-treatment and post-treatment measurement occasions are much less variable than changes within the same individuals between post-treatment and follow-up. That is, the treatment appears to have much the same initial effect on all subjects, increasing scores by about 3 units between pre-test and post-test, but there is substantial variability in change between post-test and follow-up, with some subjects returning to their pre-test level while other subjects continue to increase their scores. The multivariate model allows for the possibility that these differences in change score variances in the sample might be a reflection of similar differences in the population. The assumption of multivariate normality of Y variables implies that the difference scores are normally distributed in the population, but it implies nothing about population variances.

Confidence intervals on contrasts in planned analyses

Suppose that the experimenter has planned to construct a CI on the *within-subjects* comparison $\psi_1 = \mu_2 - \mu_1$. The population value of this comparison is also the population mean of the variable $Y_2 - Y_1$, and the standard error of $\hat{\psi}_1$ is

$$\sigma_{\hat{\psi}_1} = \frac{\sigma_{Y_2 - Y_1}}{\sqrt{n}}. \tag{6.2}$$

The ratio

$$\frac{\hat{\psi}_1 - \psi_1}{\hat{\sigma}_{\hat{\psi}_1}} = \frac{M_{Y_2 - Y_1} - \mu_{Y_2 - Y_1}}{\sqrt{\dfrac{s^2_{Y_2 - Y_1}}{n}}}$$

has a central t distribution with $(n - 1)$ degrees of freedom.

The $100(1 - \alpha)\%$ individual raw CI on ψ_1 is

$$\psi_1 = \mu_{Y_2 - Y_1} \in \left(\hat{\psi}_1 - t_{\alpha/2;\, n-1}\sqrt{\frac{s^2_{Y_2 - Y_1}}{n}},\ \hat{\psi}_1 + t_{\alpha/2;\, n-1}\sqrt{\frac{s^2_{Y_2 - Y_1}}{n}} \right).$$

For the current small data set, the CC required for a 95% individual CI is $t_{.05/2;4} = 2.7764$, so the CI half-width is $2.7764\sqrt{2/5} = 1.756$, and the CI is $\psi_1 \in (1.244,\ 4.756)$. It is of some interest to compare this CI with that on $\psi_2 = \mu_3 - \mu_2$, the difference between the *FU* and *Pre* means, which has the same sample value ($\hat{\psi}_2 = \hat{\psi}_1 = 3.0$). The half-width of the CI on ψ_2 is $2.7764\sqrt{36.5/5} = 7.502$, and the CI is $\psi_2 \in (-4.502,\ 10.502)$. This is very much wider than the CI on ψ_1, because $\hat{\psi}_2$ has a very much larger estimated standard error than $\hat{\psi}_1$, even though both contrasts are comparisons.

If SCIs are constructed in a planned analysis restricted to ψ_1 and ψ_2, the Bonferroni-t CC is $t_{.05/(2\times2);4} = 3.4954$, so the interval half-widths are $w_1 = 3.4954\sqrt{2/5} = 2.211$ and $w_2 = 3.4954\sqrt{36.5/5} = 9.444$.

Standardized within-subjects contrasts

Before a contrast can be scaled in standard deviation units it is necessary to make a decision about the basis for standardization. It is generally agreed that standardization for a between-subjects design should be based on variation within treatment populations. Given the assumption of homogeneous treatment variances $(\sigma^2_{\varepsilon_1} = \sigma^2_{\varepsilon_2} = \cdots = \sigma^2_{\varepsilon_J} = \sigma^2_{\varepsilon})$, the natural basis for standardization for a between-subjects design is the common within-population standard deviation σ_ε. There is, however, no generally accepted convention about the appropriate basis for standardization in the case of within-subjects designs (Morris and DeShon, 2002).

Unlike the univariate model for repeated measures data, the multivariate model does not require the assumption of variance homogeneity of Y scores under all treatments (or occasions of measurement). Suppose, for example, that

$$\sigma^2_{Y_1} < \sigma^2_{Y_2} < \sigma^2_{Y_3},$$

a pattern of heterogeneous population variances consistent with the pattern of sample variances in the data set we are considering. Variance heterogeneity across measurements allows for various possible approaches to standardization. The most obvious possibilities are:

- standardize all contrasts on the basis of some index of avérage variability within conditions (such as the square root of the average of the population variances);
- standardize all contrasts on the basis of pre-intervention variability (that is, variability in Y_1 scores), or variability in the 'control' condition (if a single condition has that particular status);
- standardize each contrast on the basis of variability in scores on that contrast.

The third option should probably be avoided, if only because it has no direct relationship to the kind of standardization that is now well accepted for the analysis of between-subjects designs. This option has often been used, however, to define standardized effect sizes. As Dunlap, Cortina, Vaslow and Burke (1996) point out, several published meta-analyses have incorrectly treated this type of standardized effect size as though it is commensurate with Cohen's *d*.

If standardized effect sizes from within-subjects designs are to be interpreted in the same way as standardized effect sizes from between-subjects designs, it seems clear that something like the first option should be used for routine analysis. Having to use some index of average variability in the presence of variance heterogeneity is, of course, less than ideal. The same can be said of the analogous standardization of contrasts from between-subjects designs.[2]

Assuming variance homogeneity Let us suppose for the moment that population variances on the p measurements are homogeneous, that is, that

$$\sigma^2_{Y_1} = \sigma^2_{Y_2} = \cdots = \sigma^2_{Y_p} = \sigma^2_Y.$$

Given this kind of homogeneity, σ_Y is the same kind of standard deviation as the corresponding parameter (σ_ε) defined by an ANOVA model of data from a between-subjects design. The standardized value of a contrast ψ_g from a within-subjects design, then, is ψ_g / σ_Y. If ψ_g is a mean difference contrast, then ψ_g / σ_Y is a standardized effect size commensurate with Cohen's *d*.

In the within-subjects analyses discussed in this and the following chapter, variance homogeneity will be implicitly assumed whenever standardized contrasts are discussed. If there is good reason to abandon this assumption in a particular case (as there would be if the sample variances from the current example had been obtained from a very much larger data set), then a different basis for standardization might be considered. Remember that variance homogeneity is required only to justify standard interpretations of standardized effect sizes. It is not required to justify the construction or interpretation of raw CIs derived from the multivariate (MANOVA) model.[3]

Constructing standardized confidence intervals The value of σ_Y can be estimated from

$$s_Y = \sqrt{\frac{\sum_j s^2_{Y_j}}{p}}\ .$$

An approximate standardized CI on ψ_g / σ_Y can be constructed by dividing the limits of the raw CI on ψ_g by s_Y.

For the current data set

$$s_Y = \sqrt{\frac{14.0 + 24.0 + 36.5}{3}} = 4.9833,$$

the approximate standardized sample value of ψ_1 is

$$\frac{\hat{\psi}_1}{s_Y} = \frac{3.0}{4.9833} = 0.602$$

and the approximate standardized 95% individual CI (obtained by dividing the raw interval limits of 1.244 and 4.756 by 4.9833) is

$$\frac{\psi_1}{\sigma_Y} \in (0.250, 0.954).$$

In this case the approximate standardized interval should not be taken seriously, because of the very small sample size on which it is based.[4]

Confidence intervals in post hoc analyses

As we saw in Chapter 2, the Scheffé procedure for contrasts in a between-subjects design uses a CC derived from the critical value of the ANOVA F test statistic used to test the hypothesis of homogeneity of population means. When a multivariate model is used to analyse data from a single-factor within-subjects design, the test statistic required to carry out the analogous homogeneity test is not the univariate ANOVA F statistic, but rather a multivariate test statistic known as Hotelling's T^2 (Hotelling, 1931).

Although T^2 is not the kind of statistic that one would consider computing by hand, the underlying idea is relatively straightforward. It is possible to calculate a squared t statistic for any within-subject contrast from

$$t_{\psi_g}^2 = \frac{\hat{\psi}_g^2}{\hat{\sigma}_{\hat{\psi}_g}^2}.$$

(This is also an F statistic with $\nu_1 = 1$.) Hotelling's T^2 is the maximum value of $t_{\psi_g}^2$, maximizing over all possible contrasts. That is,

$$T^2 = \left[t_{\psi_g}^2\right]_{\max}.$$

We will not deal here with the method used to calculate T^2. The important point for our purpose is that, given the assumption of multivariate normality in the population from which subjects are randomly sampled, the homogeneity hypothesis

$$H_0: \mu_1 = \mu_2 = \cdots = \mu_p$$

can be rejected if

$$T^2 > T_{\alpha; p-1, n-1}^2 = \frac{(n-1)(p-1)}{n-p+1} F_{\alpha; p-1, n-p+1} \tag{6.3a}$$

or, equivalently, if

$$\left|t_{\psi_g}\right|_{\max} > T_{\alpha; p-1, n-1} = \sqrt{\frac{(n-1)(p-1)}{n-p+1} F_{\alpha; p-1, n-p+1}}. \tag{6.3b}$$

Tables of critical values of T^2 are available in some books on multivariate analysis (such as Timm, 1975), but it is standard practice to convert T^2 into an F statistic because of the ubiquity of F tables.

The maximal contrast, meaning the contrast for which $|t| = T$, is usually of little substantive interest because of problems of interpretation. Whether or not the maximal contrast is of interest in its own right, the critical value of $\left|t_{\psi_g}\right|_{\max}$ on the right hand side of (6.3b) can be used as a CC for the construction of $100(1-\alpha)\%$ SCIs on all contrasts, including any that might be suggested by an inspection of the data.

For the current data set the CC for 95% SCIs in an unrestricted analysis is

$$\text{CC} = \sqrt{\frac{(n-1)(p-1)}{n-p+1}F_{\alpha;\, p-1,\, n-p+1}} = \sqrt{\frac{4\times 2}{3}F_{.05;\,2,\,3}} = 5.047. \qquad (6.4)$$

CI half-widths for ψ_1 and ψ_2 are

$$w_1 = 5.047\sqrt{2/5} = 3.192$$

and $$w_2 = 5.047\sqrt{36.5/5} = 13.636.$$

The width of post hoc CIs increases as the number of measurements (p) increases. In general, single-factor experiments with a large number of repeated measures (such as learning experiments with a large number of trials) should not be subjected to post hoc analysis unless the sample size is unusually large.[5]

Carrying out a planned analysis with PSY

The *PSY* input file shown below produces an ANOVA-style summary table (as well as CIs) for all comparisons. *PSY* refers to these contrasts as *W* (within-subjects) contrasts. Coefficients for within-subjects contrasts must appear under the heading [WithinContrasts] (or [WContrasts]). There are no *B* (between-subjects) contrasts, because there is only one group. Each row of the data section of the input file contains a group-membership 'variable' (with a value of 1 because there is only one group) followed by the three *Y* scores for a given subject. The input file follows.

```
[WithinContrasts]
-1  1  0 Post-Pre
 0 -1  1 FU-Post
-1  0  1 FU-Pre
[Data]
1 18 23 19
1 12 15 24
1 12 15 21
1 10 11 18
1  8 11  8
```

The Bonferroni-*t* procedure is selected from the Analysis Options window because it provides a smaller CC than the T^2 procedure. (When contrasts are planned it is perfectly legitimate to run both procedures to discover which produces narrower intervals.) An edited version of the output file (excluding standardized CIs) follows.

```
Means and Standard Deviations

        Group 1  Overall Mean:        15.000
         Measurement    1            2            3
         Mean         12.000       15.000       18.000
         SD            3.742        4.899        6.042
```

Analysis of Variance Summary Table

	Source	SS	df	MS	F
	Between	210.000	4	52.500	
	Within				
Post-Pre	W1	22.500	1	22.500	22.500
	Error	4.000	4	1.000	
FU-Post	W2	22.500	1	22.500	1.233
	Error	73.000	4	18.250	
FU-Pre	W3	90.000	1	90.000	6.545
	Error	55.000	4	13.750	

Bonferroni 95% Simultaneous Confidence Intervals

Raw CIs (scaled in Dependent Variable units)

	Contrast	Value	SE	..CI limits.. Lower	Upper
Post-Pre	W1	3.000	0.632	0.495	5.505
FU-Post	W2	3.000	2.702	-7.701	13.701
FU-Pre	W3	6.000	2.345	-3.289	15.289

It is clear from the CI table that the point estimates of the magnitudes of the comparisons W_1 and W_2 are identical, but that these two sample values have very different estimated standard errors. As a consequence, the CI on W_1 is much narrower than that on W_2. A between-subjects analysis of data from an experiment with equal sample sizes would produce identical estimated standard errors for all comparisons.

The ANOVA summary table distinguishes between two major sources of variation (Between and Within). Each contrast under the Within heading has its own error term. The difference between *Post* and *Pre* (W_1) produces a relatively large and statistically significant F ratio ($F = 22.500$, $F_c = F_{.05/3;1,4} = 15.688$), whereas the difference between *FU* and *Pre* (W_3) does not ($F = 6.545$). Those who (incorrectly) interpret F ratios as measures of effect size might be tempted to conclude that $W_1 > W_3$. The CI table shows that the sample value (and CI midpoint) of W_3 is twice as large as the sample value (and CI midpoint) of W_1, so it is quite clear that the single directional inference ($W_1 > 0$) implied by the CI analysis does not imply that $W_1 > W_3$.

Carrying out a post hoc analysis

Although it is not necessary to do so, it will be instructive to consider the results of a *PSY* post hoc analysis in conjunction with a multivariate-model homogeneity test. An excerpt from the output of a *SYSTAT* analysis follows.

```
Multivariate Repeated Measures Analysis
Test of: A                  Hypoth. df   Error df       F          P
 Wilks' Lambda=   0.0713        2            3       19.5395     0.019
 Pillai Trace =   0.9287        2            3       19.5395     0.019
 H-L Trace    =  13.0263        2            3       19.5395     0.019
```

The third test statistic is the Hotelling–Lawley trace statistic, which is not Hotelling's T^2. If required, the T^2 statistic may be calculated from the F statistic associated with all three multivariate test statistics as follows:

$$T^2 = \frac{(n-1)(p-1)}{n-p+1} F_{p-1,\, n-p+1} \tag{6.5}$$

where $F_{p-1,\, n-p+1}$ is the F value reported for the three MANOVA test statistics, all of which are equivalent in a single-group within-subjects analysis.

For the current small data set,

$$T^2 = \frac{4 \times 2}{3} \times 19.5395 = 52.105.$$

The homogeneity hypothesis can be rejected by the multivariate homogeneity test (p = .019), so a post hoc analysis must be capable of producing a directional inference on the maximal contrast. The coefficients of the maximal contrast can be obtained from a *discriminant function*, using a method outlined by Harris (1994, p.277). It turns out that the maximal contrast for this particular data set is

$$\hat{\psi}_{\max} = -913 M_1 + 750 M_2 + 163 M_3.$$

To illustrate the properties of the maximal contrast, we add the line

```
-913 750 163 Psy max
```

to the input file used for the earlier planned analysis. (In practice the maximal contrast would not usually be included in a post hoc analysis because of potential interpretation problems.) A post hoc analysis is obtained by selecting the *post hoc* option from the Analysis Options window. Edited output follows.

```
                  Analysis of Variance Summary Table
                  Source          SS        df        MS             F
                  -------------------------------------------------------
                  Between      210.000      4       52.500
                  -------------------------------------------------------
                  Within
                  -------------------------------------------------------
Post-Pre          W1            22.500      1       22.500        22.500
                  Error          4.000      4        1.000
FU-Post           W2            22.500      1       22.500         1.233
                  Error         73.000      4       18.250
FU-Pre            W3            90.000      1       90.000         6.545
                  Error         55.000      4       13.750
Psy max           W4            36.622      1       36.622        52.105
                  Error          2.811      4        0.703
                  -------------------------------------------------------
```

```
                 Post hoc 95% Simultaneous Confidence Intervals
                 ----------------------------------------------
                 The CIs refer to mean difference contrasts,
                  with coefficients rescaled if necessary.
                  The rescaled contrast coefficients are:

                 Rescaled Within contrast coefficients
           Contrast          Measurement...
                                1            2            3
Post-Pre         W1          -1.000        1.000        0.000
FU-Post          W2           0.000       -1.000        1.000
FU-Pre           W3          -1.000        0.000        1.000
Psy max          W4          -1.000        0.821        0.179

                 Raw CIs (scaled in Dependent Variable units)
                 ---------------------------------------------------
                 Contrast   Value          SE          ..CI limits..
                                                      Lower        Upper
                 ---------------------------------------------------
Post-Pre         W1          3.000        0.632       -0.192        6.192
FU-Post          W2          3.000        2.702      -10.636       16.636
FU-Pre           W3          6.000        2.345       -5.836       17.836
Psy max          W4          3.536        0.490        1.064        6.008
                 ---------------------------------------------------
```

Note that the F $(= t^2)$ value for the maximal contrast is equal to $T^2 = 52.105$, the MANOVA test statistic used for the test of homogeneity of population means, and that the post hoc CI for this contrast excludes zero, as it must when the T^2 value is statistically significant ($p = .019$ from the *SYSTAT* output).

Unlike the maximal contrast from an analysis of a between-subjects design, the maximal contrast from a multivariate-model analysis of a within-subjects design does not account for all of the variation between means. In this case, the difference between *FU* and *Pre* (W_3) accounts for all of the variation between means, with a much larger sum of squares $[\text{SS}(W_3) = 90.0]$ than that for the maximal contrast. The F ratio for W_3 is relatively small, however, because that particular contrast has a relatively large error term, due to the substantial variation between subjects in the degree of change between the first and the last of the three measurements. The differences between W_3 and the maximal contrast are reflected in the sample values of the mean difference versions of the two contrasts (W_3 has a much larger sample value) and in the estimated standard errors (the maximal contrast is estimated with much greater precision because it has a smaller estimated standard error).

Two-factor within-subjects designs

We turn now to factorial designs with no between-subjects factors. Consider an experiment concerned with the effectiveness of auditory warnings, where all subjects rate the degree of urgency in each of 12 recorded warnings. Each is a combination of levels of two factors:

A : Tone of Voice (3 levels: Urgent, Nonurgent, Monotone)
B : Message (4 levels: Deadly, Danger, Warning, Caution).

This experiment can be regarded a (3×4) two-factor design, the parentheses indicating that both factors are within-subject factors. We will henceforth adopt the convention of placing within-subjects factors in parentheses. A (3×4) design has two within-subjects factors, and each subject is measured 12 times. A $3 \times (4)$ design has one between-subjects factor with three levels (that is, three groups) and one within-subjects factor with four levels, so each subject is measured four times. We will examine *mixed* designs (that is, designs including at least one factor of each type) in Chapter 7.

The Warnings data set We will use data from an experiment including the two within-subjects factors described in the previous paragraph (Hellier, Weedon, Adams, Edworthy and Walters, 1999). Ratings of urgency were provided by $n = 31$ subjects. Each of the $p = 12$ univariate distributions of ratings was substantially positively skewed, creating a potential threat to the validity of an analysis based on the assumption of multivariate normality. Although there is evidence that ANOVA test procedures work reasonably well when applied to moderately skewed distributions (Glass, Peckham and Sanders, 1972), it was considered prudent to reduce skew by applying a monotonic (order-preserving) nonlinear transformation to the data before carrying out CI analyses of this particular data set.

A logarithmic transformation is commonly used to reduce or eliminate positive skew (Winer, Brown and Michels, 1991). The model employed for the analysis may written as

$$Y_{ijk} = \log_e(X_{ijk} + 1) = \mu_{jk} + \varepsilon_{ijk} \tag{6.6}$$

where X_{ijk} is the rating produced by subject i at level j of Factor A and level k of Factor B

$\log_e(X_{ijk} + 1)$ is the natural logarithm of $(X_{ijk} + 1)$.

The analysis is based on values of $\log_e(X + 1)$ rather than $\log_e(X)$ in order to ensure that the transformation does not change the origin. [An X value of zero is transformed into a $\log_e(X + 1)$ value of zero.]

Cell means (and standard deviations) and row and column means of transformed values are given in Table 6.1. The profile of means at each level of the Tone factor is shown in Figure 6.1. These profiles do not suggest substantial departures from additivity of Tone and Warning effects, so main effect contrasts are likely to provide a reasonable account of variation between cell means.

Table 6.1 *Means and standard deviations from Warnings experiment*

	b_1 (Deadly)	b_2 (Danger)	b_3 (Warning)	b_4 (Caution)	Row means
a_1 (Urgent)	4.580 (0.933)	4.475 (0.967)	4.370 (0.975)	4.372 (0.971)	4.449
a_2 (Nonurgent)	4.094 (0.871)	4.000 (0.944)	3.697 (0.987)	3.799 (1.009)	3.898
a_3 (Monotone)	3.936 (0.890)	3.791 (0.944)	3.672 (1.032)	3.620 (0.971)	3.755
Column means	4.204	4.088	3.913	3.930	

Analysis options

Bonferroni-*t* SCIs for planned contrasts analyses have a CC of

$$CC = t_{\alpha/(2k_{\text{fam}}); n-1} \tag{6.7}$$

where k_{fam}, the number of planned contrasts in a particular family, can vary across families or effects.

The CC for T^2 post hoc SCIs is

$$CC = \sqrt{\frac{\nu_1(n-1)}{n-\nu_1} F_{\alpha;\nu_1,n-\nu_1}} \tag{6.8}$$

where ν_1 is the number of degrees of freedom for the effect defining the family of contrasts. The value of ν_1 can vary across families. In a standard main and interaction effects analysis of a (3×4) design, ν_1 has values of $J - 1 = 2$ for the *A* main effect, $K - 1 = 3$ for the *B* main effect, and $(J - 1)(K - 1) = 6$ for the *AB* interaction effect. An analysis allowing for direct inferences on all factorial contrasts could be carried out by defining a single family with $\nu_1 = JK - 1 = 11$ and replacing α with 3α in (6.7) or with $1 - (1 - \alpha)^3$ in (6.8).

There is no generally accepted analogue of the *f* effect size parameter for multivariate-model analyses of within-subjects designs, so we will not concern ourselves with CI-based heterogeneity inference.

There is no multivariate analogue of the *SMR* procedure for post hoc analyses of $(J \times K)$ designs. [There is, however, a multivariate analogue of the *SMR* procedure for post hoc analyses of $J \times (K)$ designs, as we will see in Chapter 7.]

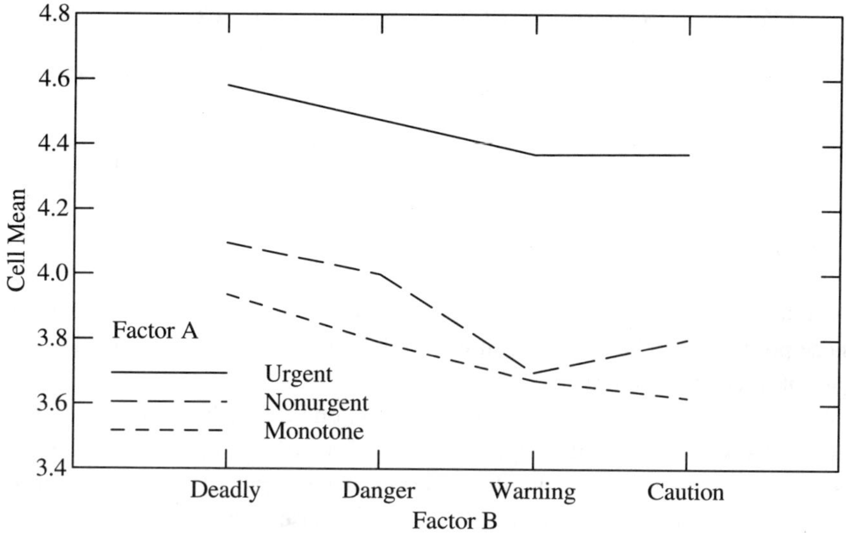

Figure 6.1 *Profiles of means from Warnings experiment*

Example 6.1 Two-factor within-subjects planned analysis

Consider the following coefficient vectors referring to factor levels.

	a_1	a_2	a_3
A_1	1	–1	0
A_2	1	0	–1
A_3	0	1	–1

	b_1	b_2	b_3	b_4
B_1	1	–1	0	0
B_2	1	0	–1	0
B_3	1	0	0	–1
B_4	0	1	–1	0
B_5	0	1	0	–1
B_6	0	0	1	–1

These coefficient vectors can be used to define main effect contrasts that are comparisons on factor levels. Provided that the decision to restrict the analysis to the main and interaction effect contrasts implied by these coefficient vectors is not influenced by an inspection of the data, the Bonferroni-*t* procedure (6.7) can be used to control the FWER in a standard factorial analysis, but it may be less efficient than the T^2 procedure (6.8) because of redundancy within the set of contrasts planned for each effect. It is therefore prudent to compare the Bonferroni-*t* CC for each effect with the corresponding T^2 CC. There is no point in imposing restrictions on the contrasts in an analysis if those restrictions do not produce an increase in precision.

Bonferroni *t* and T^2 CCs are as follows:

Effect	*Bonferroni-t CC*	T^2 *CC*
A	$t_{.05/(2\times3);30} = 2.536$	$\sqrt{\frac{2\times30}{29}F_{.05;2,29}} = 2.624$
B	$t_{.05/(2\times6);30} = 2.825$	$\sqrt{\frac{3\times30}{28}F_{.05;3,28}} = 3.078$
AB	$t_{.05/(2\times18);30} = 3.259$	$\sqrt{\frac{6\times30}{25}F_{.05;6,25}} = 4.234$

Despite the redundancy within families, the Bonferroni-*t* CCs are smaller than those produced by the T^2 procedure, and the difference is far from trivial in the case of interaction contrasts.

We will use *PSY* to carry out three analyses, one for each family of contrasts. Contrast coefficients in the input file must refer to measurements rather than factor levels. Each row of the data section of the input file contains a group-membership 'variable' (with a value of 1 because there is only one group) followed by the 12 scores for a given subject. The 12 scores are *Y* values obtained under conditions a_1b_1, a_1b_2, ..., a_3b_3, a_3b_4. Note that the *a* subscript changes more slowly than the *b* subscript.

A main effect analysis An edited version of the *PSY* input file for the *A* main effect analysis follows.

```
[WithinContrasts]
1  1  1  1   -1 -1 -1 -1    0  0  0  0 Urg-N.Urg A1
1  1  1  1    0  0  0  0   -1 -1 -1 -1 Urg-Monot A2
0  0  0  0    1  1  1  1   -1 -1 -1 -1 N.Urg-Mon A3
[Data]
1 5.77920 5.72359  .  .  .  4.36310 4.26268
1 5.36364 5.48480  .  .  .  4.29729 4.39445
.
.
1 2.58400 2.35138  .  .  .  1.55814 1.74921
```

The contrast coefficients define comparisons on levels of Factor *A*, but complex contrasts on the 12 cell means. The mean difference version of contrast A_1, for example, compares urgent with nonurgent tones, averaged across the four messages. Selection of *Bonferroni t* on the Analysis Options screen produces an analysis controlling the FWER for the *A* main effect. The *PSY* contrast labels (W1, W2 and W3) have been deleted from the following edited version of the output file. Only standardized CIs are shown here.

```
Analysis Title: A comparisons
-------------------------------------------------------------
          Bonferroni 95% Simultaneous Confidence Intervals
          ---------------------------------------------------
          The CIs refer to mean difference contrasts,
            with coefficients rescaled if necessary.
```

```
Approximate Standardized CIs (scaled in Sample SD units)
          ----------------------------------------------
          Contrast   Value      SE          ..CI limits..
                                           Lower      Upper
          ----------------------------------------------
Urg-N.Urg A1         0.575     0.055       0.435      0.716
Urg-Monot A2         0.724     0.065       0.559      0.889
N.Urg-Mon A3         0.149     0.028       0.079      0.219
          ----------------------------------------------
```

These CIs produce precise estimates of the values of the relevant comparisons, even though the sample size (n = 31) is not particularly large. The standard errors of these contrasts are impressively small because of the very high correlations (from .933 to .988) between the linear combinations of variables whose differences define the A main effect contrast variables. (These are correlations between average transformed ratings, averaging across the four levels of Factor B.) As would be expected, the correlations are inversely related to standard errors; the highest correlation of .988 (between the average ratings of nonurgent and monotone messages) is associated with the smallest standard error (of the mean difference between those average ratings).

If any difference of less than about 0.25 standard deviation units is regarded as trivially small, then the analysis justifies the conclusion that the nonurgent and monotone conditions are practically equivalent (at least as far as the Tone main effect is concerned), even though there is good evidence (from the ANOVA summary table, not shown here) that warning messages delivered in a non-urgent tone produce higher urgency ratings than monotone messages ($F_{1,30} = 29.282$, $k_A\, p = .00002$). The CI on this particular contrast [$A_3 \in (0.079\sigma, 0.219\sigma)$] provides much more information than the outcome of a significance test: the lower limit of the interval tells us that the effect is positive, while the upper limit tells us that it is trivially small. A significance test would tell us only that the effect is positive.

We can conclude from the remaining CIs that messages delivered in an urgent tone produce higher average urgency ratings (averaging across levels of Factor B) than the same messages delivered in either of the other tones, both average differences being sufficiently large to be considered nontrivial.

B main effect analysis The contrasts section of the input file for the B main effect analysis is as follows:

```
[WithinContrasts]
1 -1  0  0   1 -1  0  0   1 -1  0  0 Dead-Dang B1
1  0 -1  0   1  0 -1  0   1  0 -1  0 Dead-Warn B2
1  0  0 -1   1  0  0 -1   1  0  0 -1 Dead-Caut B3
0  1 -1  0   0  1 -1  0   0  1 -1  0 Dang-Warn B4
0  1  0 -1   0  1  0 -1   0  1  0 -1 Dang-Caut B5
0  0  1 -1   0  0  1 -1   0  0  1 -1 Warn-Caut B6
```

When the Analysis Options menu appears, we again select *Bonferroni t*. The edited output for the B main effect analysis follows.

```
Analysis Title: B comparisons
-----------------------------------------------------------

               Bonferroni 95% Simultaneous Confidence Intervals
               ---------------------------------------------
               The CIs refer to mean difference contrasts,
                 with coefficients rescaled if necessary.
          Approximate Standardized CIs (scaled in Sample SD units)
               ---------------------------------------------------
               Contrast  Value         SE           ..CI limits..
                                                  Lower       Upper
               ---------------------------------------------------
 Dead-Dang B1           0.120        0.039        0.011        0.229
 Dead-Warn B2           0.303        0.049        0.165        0.441
 Dead-Caut B3           0.285        0.053        0.136        0.434
 Dang-Warn B4           0.183        0.031        0.096        0.270
 Dang-Caut B5           0.165        0.027        0.089        0.241
 Warn-Caut B6          -0.018        0.029       -0.102        0.065
               ---------------------------------------------------
```

All of the *B* (Message) main effect comparisons are estimated with a high degree of precision. If the relative effectiveness of the four messages is evaluated in terms of *B* main effect means, then Deadly is the most effective message, although it is only slightly more effective than Danger, which is slightly more effective than both of the remaining messages. The two least effective messages (Warning and Caution) are practically equivalent. While the relations

$$\text{Deadly} > \text{Danger} > \text{Warning} \approx \text{Caution}$$

are implied by this set of intervals, the intervals also imply that differences between adjacent messages in this hierarchy are very small. The largest difference is smaller than the largest difference within the *A* main effect [$A_2 \in (0.56\sigma, 0.89\sigma)$].

AB interaction analysis Coefficients for *AB* product interaction contrasts are obtained by multiplication in the usual way. The resulting contrasts section of the input file for the analysis of interaction is as follows:

```
[WithinContrasts]
1 -1  0  0  -1  1  0  0   0  0  0  0 A1B1
1  0 -1  0  -1  0  1  0   0  0  0  0 A1B2
1  0  0 -1  -1  0  0  1   0  0  0  0 A1B3
0  1 -1  0   0 -1  1  0   0  0  0  0 A1B4
0  1  0 -1   0 -1  0  1   0  0  0  0 A1B5
0  0  1 -1   0  0 -1  1   0  0  0  0 A1B6
1 -1  0  0   0  0  0  0  -1  1  0  0 A2B1
1  0 -1  0   0  0  0  0  -1  0  1  0 A2B2
1  0  0 -1   0  0  0  0  -1  0  0  1 A2B3
0  1 -1  0   0  0  0  0   0 -1  1  0 A2B4
0  1  0 -1   0  0  0  0   0 -1  0  1 A2B5
0  0  1 -1   0  0  0  0   0  0 -1  1 A2B6
0  0  0  0   1 -1  0  0  -1  1  0  0 A3B1
0  0  0  0   1  0 -1  0  -1  0  1  0 A3B2
0  0  0  0   1  0  0 -1  -1  0  0  1 A3B3
0  0  0  0   0  1 -1  0   0 -1  1  0 A3B4
0  0  0  0   0  1  0 -1   0 -1  0  1 A3B5
0  0  0  0   0  0  1 -1   0  0 -1  1 A3B6
```

When the Analysis Options menu appears, we select *Bonferroni t* and *Interaction Contrasts*, *Within order* 1. The relevant section of output follows.

```
Analysis Title: AB contrasts
-----------------------------------------------------------
      Bonferroni 95% Simultaneous Confidence Intervals
      ----------------------------------------------------
    The coefficients are rescaled if necessary to provide
       a metric appropriate for interaction contrasts.
      Order of interaction for Within contrasts:  1

   Approximate Standardized CIs (scaled in Sample SD units)
      ----------------------------------------------------
      Contrast  Value       SE          ..CI limits..
                                        Lower       Upper
      ----------------------------------------------------
      A1B1       0.011     0.060       -0.186       0.207
      A1B2      -0.195     0.074       -0.436       0.046
      A1B3      -0.091     0.092       -0.392       0.211
      A1B4      -0.206     0.055       -0.384      -0.028
      A1B5      -0.101     0.060       -0.297       0.094
      A1B6       0.105     0.061       -0.095       0.305
      A2B1      -0.042     0.060       -0.238       0.153
      A2B2      -0.056     0.075       -0.301       0.189
      A2B3      -0.113     0.071       -0.345       0.118
      A2B4      -0.014     0.059       -0.207       0.180
      A2B5      -0.071     0.060       -0.266       0.124
      A2B6      -0.057     0.055       -0.237       0.122
      A3B1      -0.053     0.063       -0.258       0.151
      A3B2       0.139     0.065       -0.072       0.351
      A3B3      -0.023     0.052       -0.192       0.147
      A3B4       0.192     0.075       -0.051       0.435
      A3B5       0.030     0.066       -0.185       0.246
      A3B6      -0.162     0.074       -0.404       0.080
      ----------------------------------------------------
```

Each of these interaction contrasts refers to a difference between two tones in the size of a difference between two messages. The set of 18 CIs can be summarized by stating that the population values of all of these differences in differences are smaller than 0.44σ. This set of interval inferences is consistent with the possibility that Tone effects and Message effects are essentially additive, because the only interval that excludes zero $[A_1B_4 \in (-0.384\sigma, -0.028\sigma)]$ includes trivially small values. Whether or not the set of inferences *implies* the conclusion that those effects are essentially additive depends on the (implicit or explicit) definition of practical equivalence adopted by the data analyst.

Example 6.2 Two-factor within-subjects post hoc analysis

Unlike the planned analysis in Example 6.1, a post hoc analysis within each of the standard families of contrasts (*A*, *B* and *AB*) is compatible with multivariate tests of homogeneity of effects. The following excerpt from a *SYSTAT* analysis shows the results of these tests:

```
Multivariate Repeated Measures Analysis

Test of: Tone                  Hypoth. df   Error df          F           P
  Wilks' Lambda=   0.194            2          29         60.066       0.000
  Pillai Trace =   0.806            2          29         60.066       0.000
  H-L Trace    =   4.142            2          29         60.066       0.000

Test of: Message               Hypoth. df   Error df          F           P
  Wilks' Lambda=   0.358            3          28         16.746       0.000
  Pillai Trace =   0.642            3          28         16.746       0.000
  H-L Trace    =   1.794            3          28         16.746       0.000

Test of: Tone*Message          Hypoth. df   Error df          F           P
  Wilks' Lambda=   0.644            6          25          2.301       0.066
  Pillai Trace =   0.356            6          25          2.301       0.066
  H-L Trace    =   0.552            6          25          2.301       0.066
```

If required, the T^2 statistic associated with each of these tests can be calculated from

$$T^2 = \frac{\nu_1(n-1)}{\nu_2} F_{\nu_1,\nu_2}$$

where ν_1 and ν_2 are the degrees-of-freedom parameters associated with the F statistic.[6] The T^2 statistics from this analysis are T_A^2 = 124.274, T_B^2 = 53.826 and T_{AB}^2 = 16.567. The fact that T_{AB}^2 is not statistically significant by a .05-level test implies that no directional inference can emerge from 95% T^2-based SCIs on *AB* contrasts. It does not follow that the CI analysis of interaction is redundant, however, because the test outcome does not justify the inference that the interaction parameters are all zero or trivially different from zero.

When carrying out post hoc analyses of data from a factorial within-subjects design, it is necessary to remember that the *post hoc* option in *PSY* refers to the single family of all conceivable contrasts on the *p* measurement means. As a consequence, this option is not relevant for multiple-family analyses. (It can be relevant for a nonstandard *single-family* analysis allowing for the evaluation of all factorial contrasts.) The easiest way to carry out post hoc analyses within a multiple-family analysis is to

- calculate the CC for each family;
- carry out a separate analysis for each family, selecting the *User-supplied Critical Constants* option from the Analysis Options menu.

To construct post hoc SCIs we would use the T^2 CCs calculated earlier (CC_A = 2.624, CC_B = 3.078, CC_{AB} = 4.234). We will base the three analyses (one per effect) on contrasts chosen after an inspection of Table 6.1 and Figure 6.1.

A main effect analysis The contrasts section of the input file for the *A* main effect analysis follows.

```
          [WithinContrasts]
          2  2  2  2   -1 -1 -1 -1   -1 -1 -1 -1 Urg-(NU,M) A1
          0  0  0  0    1  1  1  1   -1 -1 -1 -1 N.Urg-Mon  A2
```

The choice of contrasts A_1 and A_2, based on the pattern of row means in Table 6.1, was intended to produce an *A* main effect analysis based primarily on the difference between the Urgent tone and the remaining tones. The contrast labels (A_1 and A_2) are specific to this analysis; the second contrast (A_2) in this analysis was the third contrast (A_3) in the planned *A* main effect analysis.

When the Analysis Options screen appears, we select *User-supplied Critical Constants*. Because there is only one group, only the *Within CC* option is highlighted, and we enter the required T^2 CC of 2.624.

An edited extract from the output file, excluding the ANOVA summary table and raw CIs, follows.

```
Analysis Title: A post hoc
--------------------------------------------------------------
 Special Confidence Intervals: User-supplied Critical Constants
           Within main effect CC: 2.624
           --------------------------------------------------

       Approximate Standardized CIs (scaled in Sample SD units)
           --------------------------------------------------
           Contrast  Value        SE           ..CI limits..
                                              Lower        Upper
           --------------------------------------------------
Urg-(NU,M) A1         0.650       0.059        0.495        0.804
N.Urg-Mon  A2         0.149       0.028        0.077        0.221
           --------------------------------------------------
```

Contrast A_1 (the difference between the Urgent tone and the remaining tones) accounts for most of the variation within the *A* main effect. The second contrast, orthogonal to A_1, is the difference between nonurgent and monotone tones. The CI tells us about the direction of this difference, and it also tells us that the difference is very small.

B main effect analysis The contrasts section of the input file for the *B* main effect analysis follows.

```
[WithinContrasts]
1 -1  0  0   1 -1  0  0   1 -1  0  0 Dead-Dang B1
2  0 -1 -1   2  0 -1 -1   2  0 -1 -1 Dea-(W,C) B2
0  2 -1 -1   0  2 -1 -1   0  2 -1 -1 Dan-(W,C) B3
0  0  1 -1   0  0  1 -1   0  0  1 -1 Warn-Caut B4
```

The pattern of column means in Table 6.1 suggests that the warning and caution messages may be practically equivalent, so these are combined in contrasts B_2 and B_3. Contrast B_4 is included to determine whether we can reasonably infer that these two messages are practically equivalent.

For this analysis we supply the CC 3.078. Edited output follows.

```
Analysis Title: B post hoc
--------------------------------------------------------------
 Special Confidence Intervals: User-supplied Critical Constants
           Within main effect CC: 3.078
           --------------------------------------------------
```

```
        Approximate Standardized CIs (scaled in Sample SD units)
              ---------------------------------------------------
              Contrast  Value         SE              ..CI limits..
                                                      Lower       Upper
              ---------------------------------------------------
Dead-Dang B1             0.120       0.039          0.001       0.239
Dea-(W,C) B2             0.294       0.049          0.144       0.444
Dan-(W,C) B3             0.174       0.025          0.097       0.251
Warn-Caut B4            -0.018       0.029         -0.109       0.073
              ---------------------------------------------------
```

We can conclude from the B_4 interval that, averaged across levels of the Tone factor, Warning and Caution are practically equivalent. Deadly is slightly more effective than Danger (B_1), which in turn is slightly more effective than warning–caution, the average of the practically equivalent messages (B_3). The interval around the largest contrast sample value (B_2, the difference between the deadly and warning–caution means) shows that the population value of this contrast is in the small to small–medium range.

AB interaction analysis The *AB* interaction contrasts in this analysis are product contrasts defined in the usual way. The contrasts section of the input file for the analysis of interaction follows:

```
[WithinContrasts]
2 -2  0  0   -1  1  0  0   -1  1  0  0 A1B1
4  0 -2 -2   -2  0  1  1   -2  0  1  1 A1B2
0  4 -2 -2    0 -2  1  1    0 -2  1  1 A1B3
0  0  2 -2    0  0 -1  1    0  0 -1  1 A1B4
0  0  0  0    1 -1  0  0   -1  1  0  0 A2B1
0  0  0  0    2  0 -1 -1   -2  0  1  1 A2B2
0  0  0  0    0  2 -1 -1    0 -2  1  1 A2B3
0  0  0  0    0  0  1 -1    0  0 -1  1 A2B4
```

On the Analysis Options screen we enter the CC of 4.234 and set the order of the interaction to 1. Edited output follows.

```
Analysis Title: AB post hoc

------------------------------------------------------------
 Special Confidence Intervals: User-supplied Critical Constants
           Within main effect CC: 4.234
           --------------------------------------------------
          The coefficients are rescaled if necessary to provide
             a metric appropriate for interaction contrasts.
          Order of interaction for Within contrasts:  1
       Approximate Standardized CIs (scaled in Sample SD units)
           --------------------------------------------------
           Contrast  Value         SE              ..CI limits..
                                                   Lower       Upper
           --------------------------------------------------
           A1B1     -0.016       0.051          -0.233       0.201
           A1B2     -0.114       0.069          -0.407       0.180
           A1B3     -0.098       0.041          -0.272       0.075
           A1B4      0.024       0.045          -0.166       0.214
           A2B1     -0.053       0.063          -0.319       0.213
           A2B2      0.058       0.046          -0.135       0.252
           A2B3      0.111       0.060          -0.142       0.365
           A2B4     -0.162       0.074          -0.476       0.153
           --------------------------------------------------
```

The point estimates of all contrast values are very small, which would be expected from the small departures from parallelism of the profiles of means in Figure 6.1. The CIs on some of those contrasts, however, include values that may be considered nontrivial. In particular, the CI on the difference between Non-urgent and Monotone tones in the magnitude of the difference between Warning and Caution messages (A_2B_4) includes values approaching Cohen's definition of a medium effect. Thus, while none of these intervals provides good evidence of the presence of interaction (because the value zero is included in every interval), the analysis does not establish that the population values of all eight interaction contrasts are practically equivalent to zero.

Within-subjects designs with more than two factors

No new principles are involved in CI analyses of data from multifactor within-subjects designs. If the Warning experiment had included an additional within-subject factor (*C*: Sex of speaker, with two levels), then the design would be a $(3 \times 4 \times 2)$, and every subject would have produced 24 ratings. A standard factorial analysis would define three families of main effect contrasts (*A*, *B* and *C*), three families of first-order interaction contrasts (*AB*, *AC* and *BC*) and one family of second-order interaction contrasts (*ABC*).

As is the case with between-subjects designs, it is necessary to ensure that contrast coefficients are scaled appropriately, which usually means that the sum of the positive coefficients of interaction contrasts should equal 2^x, where *x* is the order of the interaction. Critical constants for SCIs (which can vary across families) are given by (6.7) for planned analyses and (6.8) for post hoc analyses. A $(3 \times 4 \times 2)$ design has $2 \times 3 \times 1 = 6$ degrees of freedom for the second-order *ABC* interaction, so a post hoc analysis of *ABC* interaction contrasts would use a critical constant for CIs of

$$\text{CC} = \sqrt{\frac{\nu_1(n-1)}{n-\nu_1} F_{\alpha;\nu_1,n-\nu_1}} = \sqrt{\frac{6(n-1)}{n-6} F_{\alpha;6,n-6}} .$$

As is always the case, it is legitimate in a planned analysis to use the post hoc CC (6.8) if it happens to be smaller than the Bonferroni-*t* CC (6.7).

The principles discussed in Chapter 5 for the definition of families including simple effect contrasts apply to within-subjects designs as well as between-subjects designs.

Further reading

Harris (1994) gives a detailed account of multivariate-model procedures for within-subjects designs, including the procedures used to define the maximal contrast (which he calls the optimal contrast). Harris also provides a critical examination of arguments sometimes advanced in favour of the univariate-model approach. For a more technical account assuming familiarity with matrix algebra, see Harris (2001).

For discussions of the advantages and disadvantages of within-subjects designs, see Harris (1994) or Maxwell and Delaney (1990).

Questions and exercises

1. The file *SD×NVH within.in* contains the data section of a *PSY* input file with one group of 25 subjects and four measurements per subject. The *Y* values for Measurement *j* are the same (but not in the same order) as those for Group *j* in the file *SD×NVH.in*, which contains the data for the 2×2 Sleep deprivation × NVH experiment discussed in Chapter 4. As a result, the means and standard deviations are the same in both data sets. You can assume that the four measurements are from a (2×2) within-subjects design, where the factors are Sleep deprivation and NVH, and the order of measurements in the file is a_1b_1, a_1b_2, a_2b_1, a_2b_2. The dependent variable is an error score in a driving simulator.

(a) Use *PSY* to carry out a two-way ANOVA (with $\alpha = .05$) on the data in *SD×NVH within.in*. Interpret the raw CIs on main effect and interaction contrasts. You can assume that a difference of about 2.0 dependent variable units is the smallest nontrivial difference.

(b) Reanalyse the data using a procedure that allows for inferences on all factorial contrasts. What does this analysis tell you that the previous analysis did not?

2. Carry out a between-subjects analysis of the data in *SD×NVH.in* using a procedure that allows for direct inferences on all factorial contrasts. Comment on similarities and differences between the CIs from this analysis and those from the analysis you carried out in answering Question 1(b).

Notes

1. The univariate two-factor model is not the two-factor model discussed in Chapters 4 and 5, because it treats the Subjects factor as a *random* (as distinct from *fixed*) factor. See Harris (1994, Chapter 5) for a detailed critique of the univariate-model approach to the analysis of data from within-subjects designs.

2. Before Cohen's *d* became the accepted standardized effect size parameter for a Treatment vs Control comparison in a two-group between-subjects design, Glass (1976) suggested that the raw mean difference should be divided by the Control group standard

deviation. The Glass parameter is equivalent to Cohen's d if the ANOVA-model assumption of variance homogeneity ($\sigma_T^2 = \sigma_C^2$) holds.

3. Homogeneity of Y variances does not imply homogeneity of contrast variances, which depend on covariances between measurements.

4. It is not possible to construct 'exact' noncentral t-based CIs on within-subjects contrasts in the same metric as Cohen's d. If $\sigma_{Y_1} = \sigma_{Y_2} = \sigma_Y$, the statistic

$$\hat{\psi}_1 / \hat{\sigma}_{\hat{\psi}_1} = M_{Y_2 - Y_1} \Big/ \sqrt{\frac{s^2_{Y_2 - Y_1}}{n}}$$

has a noncentral t distribution with noncentrality parameter

$$\delta = \frac{\sqrt{n}\,\psi_1}{\sigma_Y \sqrt{2(1 - \rho_{Y_1 - Y_2})}}$$

where $\rho_{Y_1 - Y_2}$ is the (unknown) population correlation between Y_1 and Y_2. A standardized CI derived from a CI on δ would produce CI limits that were larger than the required limits (for a CI on Cohen's d) by a factor of $\sqrt{2(1 - \rho_{Y_1 - Y_2})}$.

5. It is usually possible to analyse experiments with a large number of trials in terms of a relatively small number of planned contrasts, as is done in trend analysis where the contrasts refer to lower-order (linear and quadratic) components of trend, or to parameters of a change or growth model. If a post hoc analysis is required, it is often sensible to carry out the analysis on a relatively small number of blocks of trials defined independently of the data, rather than on the trials themselves.

6. If you wish to report a T^2 statistic you should include the *df* parameter, which is $(n - 1)$, not the ν_2 parameter from the associated F statistic. The outcome from the AB homogeneity test can be summarized by reporting that $T^2_{30} = 16.57$ ($p = .066$).

7 Mixed Designs

In this chapter we complete our examination of factorial designs by considering mixed designs: multifactor designs with at least one between-subjects and at least one within-subjects factor. A $3 \times (4)$ design, for example, has a three-level between-subjects factor and a four-level within-subjects factor, so that each subject in each of three groups is observed four times.

Analysing mixed designs is a relatively straightforward matter provided that the distinction between the two types of factor (between- and within-subjects) is retained throughout the analysis. It must be possible to express every contrast in the analysis as a product contrast where one of the two coefficient vectors refers to groups, and the other coefficient vector refers to measurements. All factorial contrasts of interest to experimenters are likely to satisfy this requirement.

A detailed examination of a two-factor mixed design will be sufficient to illustrate all of the important statistical principles involved in the analysis of more complex mixed designs. From a statistical point of view there is only a trivial difference between a $2 \times 2 \times (2)$ and a $4 \times (2)$ design, and an equally trivial difference between a $2 \times (2 \times 2)$ and a $2 \times (4)$ design.

The social anxiety data set

Consider a $3 \times (4)$ experiment investigating a new treatment for social anxiety. The standard treatment is believed to work well in a short-term sense, but relapse over longer time periods is thought to be a problem. The new treatment includes components designed to prevent relapse. Ninety individuals seeking treatment are randomly assigned to one of three conditions: the new treatment, the standard treatment, or a minimal-contact control condition. (Those in the third condition receive the standard treatment after the experiment is concluded.) The social anxiety level of each individual is assessed on four occasions: immediately before treatment, immediately after the treatment has concluded, three months after the conclusion of treatment, and six months after the conclusion of treatment.

The design can be summarized in terms of factors and factor levels as follows:

Between-Ss factor (Treatment):

b_1: New treatment

b_2: Standard treatment

b_3: Minimal-contact control

Within-Ss factor (Measurement occasion):

w_1: Pre (Pre-treatment)

w_2: Post (Post-treatment)

w_3: FU1 (Three months follow-up)

w_4: FU2 (Six months follow-up).

We will refer to the between- and within-subjects factors as B and W respectively. (This terminology would not be used in reporting the results of an experiment like this.) Comparisons between levels of Factor B are comparisons between different groups of subjects, while comparisons between levels of Factor W reflect changes over time within the same subjects. The data set contained in the *PSY* input file *anxiety.in* was obtained from a simulation of the experiment with $n = 30$ subjects per cell ($N = 90$ subjects in all). The parameters used for the simulation (population means and error covariance matrix) are given in the *Questions and exercises* section at the end of this chapter. A profile plot of the sample means is shown in Figure 7.1. Before we consider various analyses of this data set, we will examine the model on which the analyses are based.

The multivariate means model for mixed designs

The multivariate means model for a $J \times (p)$ design may be written as

$$Y_{ijk} = \mu_{jk} + \varepsilon_{ijk} \tag{7.1}$$

where Y_{ijk} is the score on the dependent variable Y obtained by subject i in group j ($j = 1, 2, \ldots, J$) on measurement k ($k = 1, 2, \ldots, p$)

μ_{jk} is the population mean for condition j on measurement k

and ε_{ijk} is the error component associated with Y_{ijk}.

The joint distribution of error components within population j is assumed to be multivariate normal, implying that the distribution can be described completely by the *error covariance matrix* Σ_{ε_j}. It is also assumed that each of the J populations has the same error covariance matrix ($\Sigma_{\varepsilon_1} = \Sigma_{\varepsilon_2} = \cdots = \Sigma_{\varepsilon_J} = \Sigma_{\varepsilon}$). The error covariance matrix is a $p \times p$ matrix (a 4×4 matrix in the case of the social anxiety experiment), where each of the p diagonal elements is the

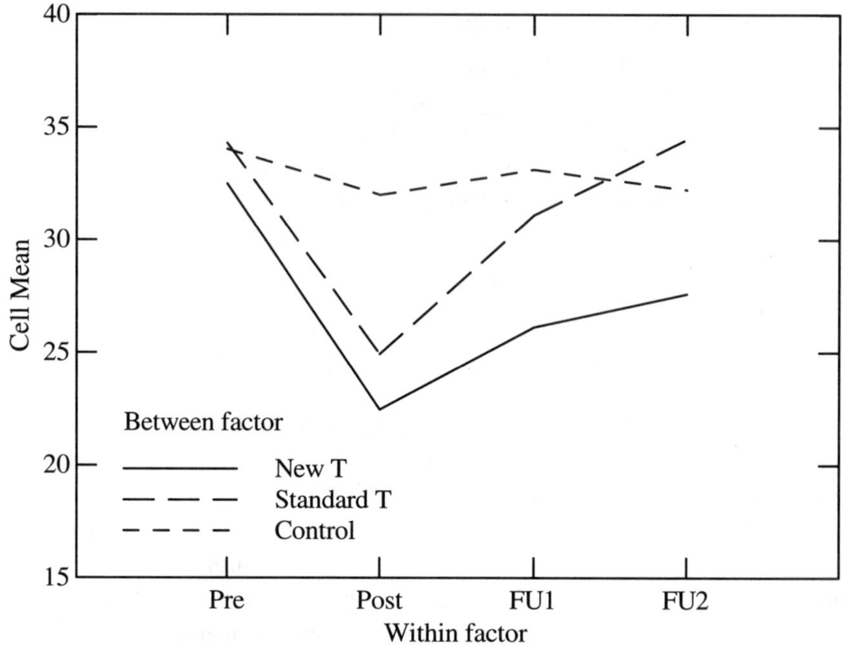

Figure 7.1 *Profile plot of means from 3 × (4) social anxiety data set*

variance of errors on one measurement occasion, and each of the off-diagonal elements is the covariance (a measure of relationship) between errors on two different measurement occasions.[1] The error covariance matrix associated with the social anxiety experiment is

$$\Sigma_\varepsilon = \begin{bmatrix} \sigma_1^2 & \sigma_{12} & \sigma_{13} & \sigma_{14} \\ \sigma_{12} & \sigma_2^2 & \sigma_{23} & \sigma_{24} \\ \sigma_{13} & \sigma_{23} & \sigma_3^2 & \sigma_{34} \\ \sigma_{14} & \sigma_{24} & \sigma_{34} & \sigma_4^2 \end{bmatrix}.$$

The assumption of homogeneity of covariance matrices implies (among other things) that the correlation between two particular measurements (such as the correlation between Pre and Post) is the same for all J populations. It does not imply that correlations between all pairs of measurements are homogeneous.

It follows from (7.1) that the distribution of Y scores within a particular population can differ from the error distribution only in the values of measurement occasion means, which must be zero for error components. As a consequence, $\Sigma_Y = \Sigma_\varepsilon$: the within-population Y score covariance matrix is identical to the error covariance matrix. While in practice Σ_ε is unknown, it can be estimated from the sample covariance matrix $\mathbf{S}_Y$, which is calculated as part of a multivariate analysis of variance. The sample within-groups covariance matrix calculated from the social anxiety data set is

$$\mathbf{S}_Y = \begin{bmatrix} 44.584 & 31.743 & 33.034 & 30.466 \\ 31.743 & 55.280 & 49.487 & 48.052 \\ 33.034 & 49.487 & 61.260 & 55.210 \\ 30.466 & 48.052 & 55.210 & 60.598 \end{bmatrix}$$

and the within-groups correlation matrix implied by $\mathbf{S}_Y$ is

$$\mathbf{R}_Y = \begin{bmatrix} 1.000 & 0.639 & 0.632 & 0.586 \\ 0.639 & 1.000 & 0.850 & 0.830 \\ 0.632 & 0.850 & 1.000 & 0.906 \\ 0.586 & 0.830 & 0.906 & 1.000 \end{bmatrix}.$$

The groups × measurements matrix of sample means from the social anxiety data set is

$$\mathbf{M} = \begin{bmatrix} 32.50 & 22.47 & 26.13 & 27.60 \\ 34.30 & 24.93 & 31.10 & 34.47 \\ 34.03 & 32.00 & 33.13 & 32.23 \end{bmatrix}.$$

Note that the number of columns of $\mathbf{M}$ is equal to the number of rows (and columns) of $\mathbf{S}_Y$. Levels of the within-subjects factor are treated as dependent variables by the multivariate model, and the decision to present the means in a 3 × 4 matrix rather than a 12 element vector is appropriate not only because we are dealing with a two-factor design, but also because the distinction between the rows and columns of $\mathbf{M}$ is required by the model. Contrasts referring to parameters of the model must be double linear combinations of means, with one coefficient vector $\mathbf{c}_B$ referring to levels of the between-subjects factor (rows of $\mathbf{M}$), and a second vector $\mathbf{c}_W$ referring to levels of the within-subjects factor (columns of $\mathbf{M}$). In matrix algebra terms, any such contrast can be defined as a product contrast of the form $\psi = \mathbf{c}'_B \boldsymbol{\mu} \mathbf{c}_W$, where $\boldsymbol{\mu}$ is the $J \times p$ matrix of population means estimated by $\mathbf{M}$. The coefficients in at least one of the vectors $\mathbf{c}_B$ or $\mathbf{c}_W$ must sum to zero.

Contrast standard errors The estimated standard error of the contrast sample value $\hat{\psi} = \mathbf{c}'_B \mathbf{M} \mathbf{c}_W$ is

$$\hat{\sigma}_{\hat{\psi}} = \sqrt{\frac{s_W^2 \sum c_B^2}{n}} \tag{7.2}$$

where s_W^2 is the sample variance of scores on the linear combination of measurements with coefficient vector $\mathbf{c}_W$

and $\sum c_B^2$ is the sum of squares of the elements in the coefficient vector $\mathbf{c}_B$ referring to groups.

The magnitude of s_W^2 is influenced by the variances and covariances in the sample covariance matrix $\mathbf{S}_Y$, as can be seen from the matrix algebra expression

$$s_W^2 = \mathbf{c}'_W \mathbf{S}_Y \mathbf{c}_W.$$

This simple-looking matrix product is equivalent to the following when $p = 4$:

$$s_W^2 = c_1^2 s_1^2 + c_2^2 s_2^2 + c_3^2 s_3^2 + c_4^2 s_4^2 + 2(c_1 c_2 s_{12} + c_1 c_3 s_{13} + c_1 c_4 s_{14} + c_2 c_3 s_{23} + c_2 c_4 s_{24} + c_3 c_4 s_{34})$$

where the coefficients are elements of the vector $\mathbf{c}'_W = [c_1 \;\; c_2 \;\; c_3 \;\; c_4]$

the s_k^2 values are measurement variances

and the $s_{kk'}$ values are measurement covariances.

Given a typical pattern of positive covariances in $\mathbf{S}_Y$ (and therefore positive correlations in $\mathbf{R}_Y$), between-subjects main effect contrasts (with uniform positive coefficients of $1/p$ in $\mathbf{c}_W$) should be based on linear combinations of measurements with relatively large values of s_W^2, so the standard errors of these contrasts should also be relatively large [from (7.2)]. Between × within interaction contrasts should have larger standard errors than within-subjects main effect contrasts, for the same reason that interaction contrasts have larger standard errors than main effect contrasts in between-subjects designs.

Example 7.1 Two-factor mixed-design planned analysis

We begin with an analysis based on a planned set of main effect and interaction contrasts using the Bonferroni-*t* procedure to construct 95% SCIs. The input file (*anxiety.in*) required to run this analysis is as follows:

```
[BetweenContrasts]
1  1 -2 Ts - C
1 -1  0 NT - ST
[WithinContrasts]
3 -1 -1 -1 Pre - rest
0  2 -1 -1 Post - FUs
0  0  1 -1 FU1 - FU2
[DATA]
1 31 26 25 31
1 44 21 27 26

(86 rows of data not shown)

3 37 27 27 30
3 25 29 26 30
```

Note that *BW* interaction contrasts are not defined in the input file. *PSY* produces appropriately scaled CIs on all *BW* contrasts derived from the *B* and *W* coefficient vectors provided in the input file. The *Bonferroni t* option is selected from the Analysis Options window. No other changes to default options are necessary.

Excerpts from the output follow.

```
Number of Groups:            3
Number of Measurements:      4

 Number of subjects in...
 Group  1:  30
 Group  2:  30
 Group  3:  30

Means and Standard Deviations
  Group 1  Overall Mean:        27.175
   Measurement   1            2             3             4
   Mean        32.500       22.467        26.133        27.600
   SD           7.624        7.001         7.031         6.360

  Group 2  Overall Mean:        31.700
   Measurement   1            2             3             4
   Mean        34.300       24.933        31.100        36.467
   SD           6.540        8.288         8.763         9.012

  Group 3  Overall Mean:        32.850
   Measurement   1            2             3             4
   Mean        34.033       32.000        33.133        32.233
   SD           5.732        6.938         7.587         7.753

 Means and SDs averaged across groups
   Measurement   1            2             3             4
   Mean        33.611       26.467        30.122        32.100
   SD           6.677        7.435         7.827         7.784

                  Analysis of Variance Summary Table
                  Source          SS       df        MS              F
                  -------------------------------------------------
                  Between
                  -------------------------------------------------
Ts - C            B1          931.612      1     931.612         5.192
NT - ST           B2         1228.538      1    1228.538         6.847
                  Error     15610.075     87     179.426
                  -------------------------------------------------
                  Within
                  -------------------------------------------------
Pre - rest        W1         1106.156      1    1106.156        42.486
                  B1W1        205.967      1     205.967         7.911
                  B2W1        148.513      1     148.513         5.704
                  Error      2265.114     87      26.036
Post -   FUs      W2         1294.252      1    1294.252       122.790
                  B1W2        470.712      1     470.712        44.658
                  B2W2        198.025      1     198.025        18.787
                  Error       917.011     87      10.540
FU1 -   FU2       W3          176.022      1     176.022        30.778
                  B1W3        186.336      1     186.336        32.581
                  B2W3        114.075      1     114.075        19.946
                  Error       497.567     87       5.719
                  -------------------------------------------------
```

Bonferroni 95% Simultaneous Confidence Intervals

The CIs refer to mean difference contrasts, with coefficients rescaled if necessary. The rescaled contrast coefficients are:

Rescaled Between contrast coefficients

	Contrast	Group...		
Ts - C	B1	0.500	0.500	-1.000
NT - ST	B2	1.000	-1.000	0.000

Rescaled Within contrast coefficients

	Contrast	Measurement...			
Pre - rest	W1	1.000	-0.333	-0.333	-0.333
Post - FUs	W2	0.000	1.000	-0.500	-0.500
FU1 - FU2	W3	0.000	0.000	1.000	-1.000

Raw CIs (scaled in Dependent Variable units)

	Contrast	Value	SE	..CI limits.. Lower	Upper
Ts - C	B1	-3.413	1.498	-6.828	0.003
NT - ST	B2	-4.525	1.729	-8.469	-0.581
Pre - rest	W1	4.048	0.621	2.532	5.564
	B1W1	3.706	1.317	0.148	7.263
	B2W1	3.633	1.521	-0.474	7.741
Post - FUs	W2	-4.644	0.419	-5.668	-3.621
	B1W2	-5.942	0.889	-8.342	-3.541
	B2W2	4.450	1.027	1.678	7.222
FU1 - FU2	W3	-1.978	0.357	-2.848	-1.108
	B1W3	-4.317	0.756	-6.358	-2.275
	B2W3	3.900	0.873	1.542	6.258

Approximate Standardized CIs (scaled in Sample SD units)

	Contrast	Value	SE	..CI limits.. Lower	Upper
Ts - C	B1	-0.458	0.201	-0.917	0.000
NT - ST	B2	-0.608	0.232	-1.138	-0.078
Pre - rest	W1	0.544	0.083	0.340	0.747
	B1W1	0.498	0.177	0.020	0.975
	B2W1	0.488	0.204	-0.064	1.040
Post - FUs	W2	-0.624	0.056	-0.761	-0.486
	B1W2	-0.798	0.119	-1.120	-0.476
	B2W2	0.598	0.138	0.225	0.970
FU1 - FU2	W3	-0.266	0.048	-0.383	-0.149
	B1W3	-0.580	0.102	-0.854	-0.306
	B2W3	0.524	0.117	0.207	0.841

Experiments of this kind are designed to examine differences between groups in the magnitude of within-subject changes over measurement occasions, so the contrasts of most interest are interaction contrasts. Main effect contrasts are of little interest. Between-subjects main effect contrasts not only have relatively large standard errors, but also refer to differences between groups averaged across measurement occasions, on the first of which there cannot possibly be systematic differences between treatments if subjects were randomly assigned to conditions. (At pre-treatment the three groups are simply different random samples from the same population.) Within-subjects main effect contrasts average across groups (including the control group), thereby addressing questions of little relevance to the concerns of the experimenter.

As expected, the average standard error of between-subject main effect contrasts (1.613 raw score units) is much larger than that of within-subject main effect contrasts (0.466). The average standard error of interaction contrasts, which are concerned with between-subject mean differences in within-subject mean differences, is approximately halfway between those two values (1.064). The average width of interaction contrast SCIs, however, is much closer to that of between-subjects main effect contrasts than to that of within-subjects main effect contrasts, due to the fact that the CC for interaction contrast SCIs ($t_{.025/(2\times6);87}$ = 2.943) is larger than the CCs for SCIs on main effect contrasts ($t_{.025/(2\times3);87}$ = 2.441 for B contrasts and $t_{.025/(2\times4);87}$ = 2.551 for W contrasts).

Directional inferences are possible for all of the interaction contrasts involving B1, the average difference between treatment and control conditions. Treatment produces a greater average reduction in social anxiety than minimal contact between the pre-treatment and subsequent measurement occasions, a greater average increase between post-treatment and the follow-ups, and a greater increase between FU1 and FU2. None of these differences is estimated precisely. It is clear from B_1W_2 that the degree of relapse among those treated (relative to the control) is in the medium to large range. Directional inferences are possible for two of the three interaction contrasts involving a comparison between the new and standard treatments, but the effects are not estimated with sufficient precision to enable useful inferences about effect sizes. It is clear from B_2W_2 that the new treatment produces less relapse than the standard treatment.

As is often the case with planned analyses, particularly planned analyses with orthogonal contrasts, this analysis does not provide answers to a number of questions that might be of interest to the experimenter, particularly questions suggested by an inspection of Figure 7.1. The analysis does not, for example, provide an inference about the differences between groups in the mean change between Pre and Post, the mean change between Pre and FU2, or the mean change between Post and FU2.

Confidence interval procedures for post hoc analyses

A traditional analysis of data from two-factor mixed designs usually begins with tests of three homogeneity hypotheses (one for each main effect and one for the interaction effect). In principle, it should be possible to replace these tests with CIs on effect size measures analogous to Cohen's f. In practice, it is possible to define and estimate an f parameter only for the between-subjects main effect, because this is the only effect concerned exclusively with between-subjects variation. No multivariate CI procedures seem to have been devised to estimate analogous parameters for the within-subjects main effect and the interaction effect. Given the fact that the viability of a post hoc contrasts analysis does not depend on access to overall effect size measures, we will not pursue this issue further. We will examine conventional homogeneity tests in order to clarify the relationship between the analyses recommended in this book and those typically reported in the research literature. We begin, however, with a discussion of post hoc SCIs on contrasts.

B main effect contrasts The CC for Scheffé SCIs on between-subjects main effect contrasts is

$$\mathrm{CC}_B = \sqrt{(J-1)\,F_{\alpha;J-1,N-J}}\,. \tag{7.3}$$

The critical F value $\left(F_{\alpha;J-1,N-J}\right)$ in (7.3) is the critical value associated with the standard univariate ANOVA F test of the hypothesis of homogeneity of the J population means of the average scores obtained on the p measurements. This is a univariate rather than a multivariate test statistic because it is not concerned with variation within subjects across measurements.

W main effect contrasts Within-subjects main effect SCIs make use of a critical value of the multivariate T^2 statistic used to test the hypothesis of homogeneity of the p measurement means, averaged across the J populations. This hypothesis is rejected if

$$T^2 > T^2_{\alpha;p-1,N-J} = \frac{(N-J)(p-1)}{N-J-p+2}\,F_{\alpha;p-1,N-J-p+2}\,. \tag{7.4}$$

When $J = 1$, the T^2 critical value specializes to the multivariate critical value (6.3a) discussed in Chapter 6. The CC associated with (7.4) is

$$\mathrm{CC}_W = \sqrt{\frac{(N-J)(p-1)}{N-J-p+2}\,F_{\alpha;p-1,N-J-p+2}}\,. \tag{7.5}$$

BW product interaction contrasts The CC required to control the FWER for *BW* product interaction contrasts uses a critical value from a largest root distribution analogous to the *SMR* distribution discussed in Chapter 5. The *GCR* (greatest characteristic root) distribution (Roy, 1953) required for the current application is somewhat more complex than the *SMR* distribution, because it takes into account the fact that each within-subjects contrast has its own error term.

Given the data from a two-factor mixed design, it is possible to determine the two contrast coefficient vectors $\mathbf{c}_{B_m}$ and $\mathbf{c}_{W_m}$ that define the *maximal product interaction contrast*

$$\hat{\psi}_m = \mathbf{c}'_{B_m}\mathbf{M}\,\mathbf{c}_{W_m}\,. \tag{7.6}$$

The statistic maximized by this interaction contrast is

$$t_{\psi}^2 = \frac{\hat{\psi}^2}{\hat{\sigma}_{\hat{\psi}}^2},$$

the contrast F statistic given in *PSY* output. If all interaction contrasts have a population value of zero, then the sampling distribution of this maximal t^2 (or maximal contrast F) statistic is the distribution of

$$\frac{(N - J)\theta_{s,m,n}}{1 - \theta_{s,m,n}},$$

where $\theta_{s,m,n}$ is the *GCR* distribution with parameters

$$s = \min(J - 1,\ p - 1), \quad m = \frac{|J - p| - 1}{2} \quad \text{and} \quad n = \frac{N - J - p}{2}.$$

It follows that 100(1 – α)% SCIs on all product interaction contrasts, including that with the maximal value of $F = t^2$, can be constructed by setting the CC at

$$\mathrm{CC}_{BW} = \sqrt{\frac{(N - J)\theta_{\alpha;s,m,n}}{1 - \theta_{\alpha;s,m,n}}}. \tag{7.7}$$

Critical values of the *GCR* distribution can be obtained from tables in Harris (2001), or from the *PSY* Probability Calculator if $s > 1$. If $s = 1$, the *GCR* CC can be expressed as an F-based CC, either (7.5) if $s = 1$ because $(J - 1) = 1$, or a Scheffé CC (7.3) if $s = 1$ because $(p - 1) = 1$.

If you are using *PSY* to carry out a CI analysis of data from a two-factor mixed design, you can produce *GCR* CIs on interaction contrasts by selecting the *post hoc* option from the Analysis Options menu. [The post hoc option uses (7.3) for *B* main effect contrasts, (7.5) for *W* main effect contrasts and (7.7) for *BW* interaction contrasts.]

Example 7.2 Two-factor mixed-design post hoc analysis

The following CI analysis is compatible with a multivariate-model two-way ANOVA producing an overall (homogeneity) test for each main effect and the interaction effect. We will discuss the homogeneity tests after the CI analysis.

The analysis includes the contrasts discussed in the previous analysis, but adds contrasts suggested by an inspection of Figure 7.1. The input file includes four *B* and six *W* contrast coefficient vectors. There is no restriction on the number of coefficient vectors, and there is also no reason why the experimenter should not subsequently run further analyses to produce inferences on additional main effect and interaction contrasts if it seems appropriate to do so. The cost of this flexibility is an increase in critical values and therefore CI width, relative to the restricted (planned) analysis discussed earlier (Example 7.1).

In a *PSY* analysis, all of the relevant post hoc CI procedures are implemented by selecting *post hoc* from the Confidence Interval menu in the Analysis Options window. The following excerpts from the output file (which is somewhat lengthy) include standardized but not raw CIs.

```
Between contrast coefficients
        Contrast      Group...
                       1    2     3
 Ts-C           B1     1    1    -2
 NT-ST          B2     1   -1     0
 NT-C           B3     1    0    -1
 ST-C           B4     0    1    -1
Within contrast coefficients
        Contrast      Measurement...
                       1    2     3    4
 Pre-rest       W1     3   -1    -1   -1
 Post-FUs       W2     0    2    -1   -1
 FU1-FU2        W3     0    0     1   -1
 Pre-Post       W4     1   -1     0    0
 Post-FU2       W5     0    1     0   -1
 Pre-FU2        W6     1    0     0   -1

             ------------------------------------------------
             Post hoc 95% Simultaneous Confidence Intervals
             ----------------------------------------------
             The CIs refer to mean difference contrasts,
              with coefficients rescaled if necessary.
              The rescaled contrast coefficients are:
             Rescaled Between contrast coefficients
       Contrast           Group...
Ts-C            B1          0.500        0.500       -1.000
NT-ST           B2          1.000       -1.000        0.000
NT-C            B3          1.000        0.000       -1.000
ST-C            B4          0.000        1.000       -1.000
             Rescaled Within contrast coefficients
       Contrast           Measurement...
Pre-rest        W1          1.000       -0.333       -0.333       -0.333
Post-FUs        W2          0.000        1.000       -0.500       -0.500
FU1-FU2         W3          0.000        0.000        1.000       -1.000
Pre-Post        W4          1.000       -1.000        0.000        0.000
Post-FU2        W5          0.000        1.000        0.000       -1.000
Pre-FU2         W6          1.000        0.000        0.000       -1.000
             ---------------------------------------------------

         Approximate Standardized CIs (scaled in Sample SD units)
             ---------------------------------------------------
             Contrast     Value         SE          ..CI limits..
                                                    Lower       Upper
             ---------------------------------------------------
Ts-C         B1          -0.458        0.201       -0.959       0.043
NT-ST        B2          -0.608        0.232       -1.186      -0.029
NT-C         B3          -0.762        0.232       -1.341      -0.184
ST-C         B4          -0.154        0.232       -0.733       0.424
Pre-rest     W1           0.544        0.083        0.303       0.784
             B1W1         0.498        0.177       -0.104       1.100
             B2W1         0.488        0.204       -0.207       1.183
             B3W1         0.742        0.204        0.047       1.437
             B4W1         0.254        0.204       -0.441       0.949
```

Post-FUs	W2	-0.624	0.056	-0.786	-0.461
	B1W2	-0.798	0.119	-1.204	-0.392
	B2W2	0.598	0.138	0.129	1.067
	B3W2	-0.499	0.138	-0.968	-0.030
	B4W2	-1.097	0.138	-1.566	-0.628
FU1-FU2	W3	-0.266	0.048	-0.404	-0.127
	B1W3	-0.580	0.102	-0.925	-0.234
	B2W3	0.524	0.117	0.125	0.923
	B3W3	-0.318	0.117	-0.717	0.081
	B4W3	-0.842	0.117	-1.241	-0.443
Pre-Post	W4	0.960	0.085	0.713	1.206
	B1W4	1.030	0.181	0.413	1.646
	B2W4	0.090	0.209	-0.622	0.801
	B3W4	1.075	0.209	0.363	1.786
	B4W4	0.985	0.209	0.273	1.697
Post-FU2	W5	-0.757	0.063	-0.938	-0.575
	B1W5	-1.088	0.134	-1.542	-0.634
	B2W5	0.860	0.154	0.335	1.384
	B3W5	-0.658	0.154	-1.183	-0.133
	B4W5	-1.518	0.154	-2.042	-0.993
Pre-FU2	W6	0.203	0.094	-0.069	0.475
	B1W6	-0.058	0.200	-0.738	0.621
	B2W6	0.949	0.231	0.164	1.734
	B3W6	0.416	0.231	-0.368	1.201
	B4W6	-0.533	0.231	-1.318	0.252

There is, of course, a great deal of redundancy in this set of contrasts. We can feel free in this post hoc analysis to base our account of the interaction on any interpretable subset of these contrasts that seems to do justice to the pattern of means shown in Figure 7.1.

Given the profiles of means in Figure 7.1, interaction contrasts involving changes between Pre and Post (W_4) and Post and FU2 (W_5) provide a straightforward account of departures from parallelism, at least at the level of directional inference. Both treatments produce a greater reduction in social anxiety between Pre and Post than does minimal contact (B_3W_4 and B_4W_4). Although these differences are estimated with poor precision, the CI on B_1W_4 suggests that the average of the treatment effects is unlikely to be trivial and may well be substantial. While there is no suggestion in the data of a difference between treatments in the magnitude of this reduction, the CI on B_2W_4 is too wide to exclude the possibility of a medium–large difference in either direction.

The magnitude of relapse between Post and FU2 is greater for the standard treatment than for the new treatment (B_2W_5) by at least a third of a standard deviation, and may be much larger. The new treatment produces more relapse than minimal contact (B_3W_5), but it is not clear whether this difference is substantial. The difference in relapse between the standard treatment and minimal contact (B_4W_5) is clearly large or very large according to Cohen's effect size guidelines.

Tests of homogeneity hypotheses

The CI table from the unrestricted two-way analysis includes at least one directional inference from each of the three contrast families (*B* main effect, *W* main effect and *BW* interaction). It follows that the three corresponding homogeneity tests (redundant in this case) must produce heterogeneity inferences, or, in the language more commonly used in practice, both main effects and the interaction effect must be statistically significant according to the relevant overall tests.

A homogeneity test is not redundant if none of the contrasts examined is statistically significant according to the relevant post hoc CI procedure. If the homogeneity hypothesis is not rejected by the relevant test, then the experimenter knows that no conceivable contrast can be declared nonzero in an unrestricted analysis. If the homogeneity hypothesis is rejected, then the maximal contrast within the relevant family (and almost certainly some additional similar contrasts) can be declared nonzero.

The main reason for discussing homogeneity tests here is to show the relationship between the analyses recommended in this book and the analyses typically reported in the research literature. We base the discussion on (edited) *SYSTAT* output from a standard repeated measures analysis of the social anxiety data set. *SYSTAT* was chosen for this analysis partly because it provides an exact *GCR* test, and partly because it provides the output in a compact form.

```
------------------------------------------------------------------
Univariate and Multivariate Repeated Measures Analysis
Between Subjects
----------------

Source              SS         df        MS            F            P

TREATMENT        2160.150       2     1080.075        6.020        0.004
Error           15610.075      87      179.426

Within Subjects
---------------

Source              SS      df       MS        F        P      G-G     H-F

MEASUREMENT     2576.431     3    858.810   60.915   0.000   0.000   0.000

MEASUREMENT
*TREATMENT      1323.628     6    220.605   15.647   0.000   0.000   0.000
Error           3679.692   261     14.098

Greenhouse-Geisser Epsilon:         0.7088
Huynh-Feldt Epsilon       :         0.7438
------------------------------------------------------------------
Multivariate Repeated Measures Analysis

Test of: MEASUREMENT         Hypoth. df   Error df      F           P
  Wilks' Lambda=       0.279       3          85      73.160      0.000
  Pillai Trace =       0.721       3          85      73.160      0.000
  H-L Trace    =       2.582       3          85      73.160      0.000
```

```
Test of: MEASUREMENT          Hypoth. df   Error df      F            P
         *TREATMENT
  Wilks' Lambda=       0.374        6          170       18.007       0.000
  Pillai Trace =       0.699        6          172       15.386       0.000
  H-L Trace    =       1.482        6          168       20.742       0.000
  Theta        =    0.572 S =    2, M = 0.0, N = 41.5 P =     0.000
----------------------------------------------------------------------------
```

As the heading indicates, *SYSTAT* produces both a univariate-model and a multivariate-model analysis. With the exception of the analysis of the Between (Treatment) main effect, for which the univariate and multivariate models produce the same analysis, the univariate section of the output is irrelevant for our purposes. It is important to be able to recognize a univariate analysis of repeated measures data, however, because researchers often report analyses of this kind without making it clear which model has been used. Two points should be noted:

- The univariate model produces a single error term for within-subjects tests based on $(N - J)(p - 1) = 87 \times 3 = 261$ degrees of freedom. Any analysis with more than N degrees of freedom for error is based on the univariate model.
- Any analysis that refers to the Greenhouse–Geisser or Huynh–Feldt epsilon statistic is based on the univariate model. These statistics are used to correct the tendency of univariate homogeneity tests to produce inflated error rates when the *sphericity* assumption fails.[2] They do not produce separate error terms for different within-subjects contrasts, and therefore do not address the main problem with univariate-model analyses. See Harris (1994, p.292) for a detailed critique of epsilon-corrected univariate tests.

B main effect test The test of the Between (Treatment) main effect ($F_{2,87} = 6.020$, $p = .004$) justifies the inference that the three treatments have heterogeneous population means, averaged across measurement occasions. It must therefore be possible to define at least one B main effect contrast with a Scheffé CI excluding zero. This inference is, of course, redundant given the results of the CI analysis.

W main effect test The multivariate test of the Within (Measurement occasion) main effect justifies the conclusion that the four repeated measurements have heterogeneous population means, averaged across treatments. The three multivariate test statistics given for the Measurement main effect are equivalent to each other (and to the T^2 statistic), and can all be converted to the same F statistic ($F_{p-1, N-J-p+2} = F_{3,85} = 73.160$, $p < .001$). The T^2 test statistic (not given in the output) is

$$T^2 = \frac{(N-J)(p-1)}{N-J-p+2} F_{p-1, N-J-p+2} = \frac{87 \times 3 \times 73.16}{85} = 224.643.$$

Therefore the maximal W main effect contrast has a $t^2_{87} = F_{1,87}$ value of 224.643. It turns out that the mean difference version of this maximal contrast has a coefficient vector of $\mathbf{c}'_W$ = [0.372 –0.861 –0.139 0.628]. (These coefficients were obtained from a discriminant function analysis. If you are interested in the computational details, consult Harris, 1994, pp. 324-337.) A *PSY* analysis shows that the standardized CI for this maximal contrast is (0.617, 0.911). Although it is not readily interpretable, the maximal contrast is reasonably similar to the difference between FU2 and Post ($-W_5$), the contrast with the largest contrast t^2 value ($t^2_{87} = F_{1,87} = 144.44$) in the CI analysis.

BW interaction test The final section of the *SYSTAT* output provides four different multivariate test statistics which can be used to test the hypothesis of no interaction between the Between (Treatment) and Within (Measurement occasion) factors. Of these, Theta ($\theta = 0.572$) is the only statistic relevant for our purposes, because this is the *GCR* statistic whose critical value is required to construct post hoc SCIs on product interaction contrasts using (7.7). The *GCR* parameters $s = \min(J-1, p-1) = 2$, $m = \left(|J-p|-1\right)/2 = 0$ and $n = (N-J-p)/2 = 41.5$ are provided in the *SYSTAT* output. When $\alpha = .05$, the critical value of the *GCR* statistic (obtained from the *PSY* probability calculator) is $\theta_{.05;2,0,41.5} = 0.119$.

If $\theta = 0.572$, the maximal product interaction contrast must have a t^2 value of

$$\frac{(N-J)\theta}{1-\theta} = 116.300.$$

A discriminant function analysis shows that this contrast has coefficient vectors of $\mathbf{c}'_B$ = [0.012 0.988 –1.000] and $\mathbf{c}'_W$ = [0.164 –0.645 –0.355 0.836], and a *PSY* analysis shows that the *GCR* 95% standardized CI on the contrast is (0.815, 1.552). We might expect a similar contrast with simplified coefficient vectors (such as $-B_4W_5$ where $\mathbf{c}'_{B_4}$ = [0 1 –1] and $\mathbf{c}'_{W_5}$ = [0 1 0 –1]) to have a similar (but necessarily smaller) t^2 value, and a similar post hoc 95% CI. It turns out that $-B_4W_5$ has a larger t^2 value (96.864) than any of the remaining 23 interaction contrasts examined in the post hoc analysis. The standardized CI on $-B_4W_5$ is (0.993, 2.042), noticeably wider than that on the maximal product interaction contrast, but with a higher midpoint (1.518 compared with the maximal contrast's 1.183). If you find it surprising that the maximal contrast can have a lower sample value (CI midpoint) than a similar contrast with a lower t^2 value, remember that the quantity maximized is the ratio of the squared sample value to the squared standard error, not the sample value (or the squared sample value) itself.

In this particular analysis the yield from these rather complex multivariate tests is minimal, given the unrestricted CI analysis already carried out. It would have been possible, of course, to carry out the tests before examining individual contrasts, with a view to discovering whether a search for statistically significant contrasts might be a waste of time. As we have seen in a number of analyses, however, CIs can be informative even when directional inference is not possible, because of the information they provide about the precision with which contrast values are estimated. The multivariate procedures are useful primarily because they provide CCs for the construction of unrestricted SCIs on *W* main effect and *BW* product interaction contrasts, not because they provide homogeneity tests.

Alternative multivariate test statistics

Before leaving the *SYSTAT* output we should perhaps briefly consider the remaining three multivariate test statistics, particularly since one of these (Pillai's trace) is often recommended in preference to Roy's *GCR* statistic as a basis for carrying out MANOVA analyses. While it can be argued that the Pillai trace statistic is usually the most appropriate MANOVA test statistic for the purpose of carrying out a homogeneity test (Olson, 1974, 1976), Roy's *GCR* statistic has no serious competitor for the purpose of constructing SCIs on product interaction contrasts, because *GCR* CCs for product contrasts are always smaller than CCs derived from critical values of any other MANOVA test statistic. The relationship between the *GCR* and Pillai CCs is directly analogous to the relationship between *SMR* and *F* (Scheffé) CCs for product interaction contrasts in two-factor between-subject designs. In both cases the relevant largest root distribution (*GCR* or *SMR*) produces a CC based on the sampling distribution of the maximum value of a product interaction contrast statistic, and therefore produces nonconservative SCIs for the family of all product interaction contrasts. In both cases a CC derived from the generally recommended test statistic (Pillai's trace or *F*) produces conservative SCIs on product interaction contrasts, and is therefore less appropriate than the relevant largest root test statistic for an analysis of product interaction contrasts.

The remaining multivariate test statistics (Wilks' lambda and the Hotelling–Lawley trace) also provide unnecessarily conservative CCs for SCIs on product contrasts. In practice, researchers who carry out homogeneity tests with MANOVA test statistics other than the *GCR* statistic usually adopt a sequential hypothesis testing approach, carrying out liberal univariate 'follow-up' tests on simple effects if (and only if) the initial multivariate test rejects the interaction homogeneity hypothesis. Incoherent analyses of this kind do not control familywise error rates, and they do not provide access to CIs.

Allowing for inferences on simple effect contrasts

Provision can be made for inferences on all factorial contrasts of interest (including simple effect contrasts) by defining one or more families of contrasts in such a way that every contrast of interest is included in one of the families so defined. Variation between cell means in two-factor mixed designs can be partitioned in a number of ways, but we will restrict our attention here to partitions analogous to those discussed in Chapter 5 in the context of between-subjects factorial designs. We will not deal explicitly with non-standard planned analyses with Bonferroni-*t* CCs, because no new principles are involved. CCs for the three most obvious analysis strategies allowing for the evaluation of post hoc contrasts are given in Table 7.1.[3]

Each of the analyses referred to in Table 7.1 defines one family which includes *BW* product interaction contrasts, and which therefore requires a CC based on a critical value of a *GCR* distribution. It is important to note that the parameters of these distributions (s, m and n) depend on what other types of *BW* product contrasts are included in the relevant family. Values of the required *GCR* parameters can be obtained from the following expressions:

Family	s	m	n
$B(W)$	$\min[(J-1),\ p]$	$\dfrac{\lvert J-p-1\rvert-1}{2}$	$\dfrac{N-J-p-1}{2}$
$W(B)$	$\min[J,\ (p-1)]$	$\dfrac{\lvert J-p+1\rvert-1}{2}$	$\dfrac{N-J-p}{2}$
All	$\min(J,\ p)$	$\dfrac{\lvert J-p\rvert-1}{2}$	$\dfrac{N-J-p-1}{2}$

CCs for various analyses of the social anxiety data set are given in Table 7.2. To carry out a nonstandard analysis, obtain the critical value from the Probability Calculator, calculate the required CC from the relevant expression in Table 7.1, and enter this as a user-supplied CC in the analysis. In the case of a nonstandard analysis with two families [$B(W)+W$ or $W(B)+B$], the primary family [$B(W)$ or $W(B)$] is likely to be the only family of interest. If so, the analysis can be carried out with a single run through *PSY*. The default scaling option produces appropriate scaling for all contrasts (including *BW* interaction contrasts), so it is not necessary to carry out a separate analysis of interaction contrasts.

A standard or nonstandard analysis of data from a two-factor mixed design is easier to carry out with *PSY* than a two-factor between-subjects or within-subjects analysis because the program recognizes the distinction between *B* and *W* factors, but not the distinction between two *B* factors or two *W* factors.

Table 7.1 *Critical constants for unrestricted analyses of data from mixed designs*

Analysis	Partition	Sub-effects	Critical constant[1]
B(W) + W	*B(W)*		$\sqrt{\frac{(N-J)\theta_{1-(1-\alpha)^2;\,s,m,n}}{1-\theta_{1-(1-\alpha)^2;\,s,m,n}}}$
		$B(w_k)$	
		B	
		BW	
	W		$\sqrt{\frac{(N-J)(p-1)}{N-J-p+2}F_{\alpha;\,p-1,\,N-J-p+2}}$
W(B) + B	*W(B)*		$\sqrt{\frac{(N-J)\theta_{1-(1-\alpha)^2;\,s,m,n}}{1-\theta_{1-(1-\alpha)^2;\,s,m,n}}}$
		$W(b_j)$	
		W	
		BW	
	B		$\sqrt{(J-1)F_{\alpha;(J-1),\,N-J}}$
All factorial effects	–		$\sqrt{\frac{(N-J)\theta_{1-(1-\alpha)^3;\,s,m,n}}{1-\theta_{1-(1-\alpha)^3;\,s,m,n}}}$
		$B(w_k)$	
		$W(b_j)$	
		B	
		W	
		BW	

[1] Values of s, m and n vary across analyses. See text.

Table 7.2 *Critical constants for unrestricted analysis of social anxiety data set*

Analysis	Partition	Critical constant
Standard	B	$\sqrt{2F_{.05;2,87}} = 2.491$
	W	$\sqrt{\frac{87\times 3}{85}F_{.05;3,85}} = 2.886$
	BW	$\sqrt{\frac{87\,\theta_{.05;2,0,41.5}}{1-\theta_{.05;2,0,41.5}}} = 3.402$
$B(W) + W$	$B(W)$	$\sqrt{\frac{87\,\theta_{.0975;2,0.5,41}}{1-\theta_{.0975;2,0.5,41}}} = 3.438$
	W	$\sqrt{\frac{87\times 3}{85}F_{.05;3,85}} = 2.886$
$W(B) + B$	$W(B)$	$\sqrt{\frac{87\,\theta_{.0975;3,-0.5,41.5}}{1-\theta_{.0975;3,-0.5,41.5}}} = 3.487$
	B	$\sqrt{2F_{.05;2,87}} = 2.491$
All factorial effects	–	$\sqrt{\frac{87\,\theta_{.1426;3,0,41}}{1-\theta_{.1426;3,0,41}}} = 3.641$

Example 7.3 GCR-based SCIs on all factorial contrasts

Table 7.2 shows that an analysis of the social anxiety data allowing for inferences on all factorial contrasts can be carried out by using a CC of 3.641 for all SCIs. Note that although this is a much larger CC than that used for B or W main effect contrasts in the standard analysis (2.491 and 2.886 respectively), it is not very much larger than the CC used in that analysis for SCIs on interaction contrasts (3.402). Given that main effect contrasts are of little interest in the case of this particular experiment, the experimenter may well regard the slight loss of precision in estimating interaction contrasts as a small price to pay for the possibility of making direct inferences on all factorial contrasts of interest.

The following analysis is produced when a user-supplied CC of 3.641 is used to construct all SCIs on the contrasts defined in the input file *anxiety all.in.* Coefficient vectors B5, B6, B7, B8, W4 and W5 are not contrast coefficient vectors and therefore produce a warning in the output file. These coefficient vectors are included to provide the basis for the definition of within-subjects simple effect or subset effect contrasts. After rescaling, for example, B5 averages across the new and standard treatments, so the subset effect contrast B5W1 is the difference between Pre and Post averaged across those treatments $[W_1((b_1+b_2)/2)]$. Some rows of the *PSY* output file (namely the 'main effects' for B5, B6, B7, B8, W4 and W5, and 'interaction contrasts' involving two such vectors, such as B5W4) refer to double linear combinations that are not contrasts, because neither of the coefficient vectors is a contrast coefficient vector. There is no real risk of mistaking irrelevant double linear combinations for contrasts, mainly because of obvious interpretation problems. Statistics referring to irrelevant double linear combinations have been edited out of the following excerpt from the output file.

```
Between contrast coefficients
          Contrast      Group...
                         1    2    3
  Ts-C            B1     1    1   -2
  NT-ST           B2     1   -1    0
  NT-C            B3     1    0   -1
  ST-C            B4     0    1   -1
  Ts              B5     1    1    0
  NT              B6     1    0    0
  ST              B7     0    1    0
  C               B8     0    0    1

  *** Caution ***
  B5  coefficients do not sum to zero
  B6  coefficients do not sum to zero
  B7  coefficients do not sum to zero
  B8  coefficients do not sum to zero
Within contrast coefficients
          Contrast      Measurement...
                         1    2    3    4
  Pre-Post        W1     1   -1    0    0
  Post-FU2        W2     0    1    0   -1
  Pre-FU2         W3     1    0    0   -1
  Post            W4     0    1    0    0
  FU2             W5     0    0    0    1

  *** Caution ***
  W4  coefficients do not sum to zero
  W5  coefficients do not sum to zero
    Special Confidence Intervals: user-supplied Critical Constants
               Between main effect CC: 3.641
               Within main effect CC: 3.641
               Between x Within interaction CC: 3.641
             -----------------------------------------------------
              The CIs refer to mean difference contrasts,
               with coefficients rescaled if necessary.
```

The rescaled contrast coefficients are:

Rescaled Between contrast coefficients

	Groups			
Ts-C	B1	0.500	0.500	-1.000
NT-ST	B2	1.000	-1.000	0.000
NT-C	B3	1.000	0.000	-1.000
ST-C	B4	0.000	1.000	-1.000
Ts	B5	0.500	0.500	0.000
NT	B6	1.000	0.000	0.000
ST	B7	0.000	1.000	0.000
C	B8	0.000	0.000	1.000

Rescaled Within contrast coefficients

	Measurements				
Pre-Post	W1	1.000	-1.000	0.000	0.000
Post-FU2	W2	0.000	1.000	0.000	-1.000
Pre-FU2	W3	1.000	0.000	0.000	-1.000
Post	W4	0.000	1.000	0.000	0.000
FU2	W5	0.000	0.000	0.000	1.000

Approximate Standardized CIs (scaled in Sample SD units)

	Contrast	Value	SE	..CI limits.. Lower	Upper
Ts-C	B1	-0.458	0.201	-1.191	0.274
NT-ST	B2	-0.608	0.232	-1.453	0.238
NT-C	B3	-0.762	0.232	-1.608	0.083
ST-C	B4	-0.154	0.232	-1.000	0.691
Ts	B5				
NT	B6				
ST	B7				
C	B8				
Pre-Post	W1	0.960	0.085	0.649	1.271
	B1W1	1.030	0.181	0.370	1.689
	B2W1	0.090	0.209	-0.672	0.851
	B3W1	1.075	0.209	0.313	1.836
	B4W1	0.985	0.209	0.223	1.747
	B5W1	1.303	0.105	0.922	1.684
	B6W1	1.348	0.148	0.809	1.886
	B7W1	1.258	0.148	0.720	1.797
	B8W1	0.273	0.148	-0.265	0.812
Post-FU2	W2	-0.757	0.063	-0.986	-0.527
	B1W2	-1.088	0.134	-1.574	-0.602
	B2W2	0.860	0.154	0.298	1.421
	B3W2	-0.658	0.154	-1.220	-0.097
	B4W2	-1.518	0.154	-2.079	-0.956
	B5W2	-1.119	0.077	-1.400	-0.839
	B6W2	-0.689	0.109	-1.087	-0.292
	B7W2	-1.549	0.109	-1.946	-1.152
	B8W2	-0.031	0.109	-0.428	0.366
Pre-FU2	W3	0.203	0.094	-0.140	0.546
	B1W3	-0.058	0.200	-0.786	0.669
	B2W3	0.949	0.231	0.109	1.789
	B3W3	0.416	0.231	-0.424	1.256
	B4W3	-0.533	0.231	-1.373	0.307
	B5W3	0.184	0.115	-0.236	0.604
	B6W3	0.658	0.163	0.064	1.252
	B7W3	-0.291	0.163	-0.885	0.303
	B8W3	0.242	0.163	-0.352	0.836
Post	W4				
	B1W4	-1.115	0.223	-1.928	-0.302
	B2W4	-0.331	0.258	-1.270	0.608
	B3W4	-1.280	0.258	-2.219	-0.342
	B4W4	-0.949	0.258	-1.888	-0.010

```
FU2            W5
               B1W5      -0.027      0.234      -0.878      0.824
               B2W5      -1.191      0.270      -2.174     -0.208
               B3W5      -0.622      0.270      -1.605      0.361
               B4W5       0.569      0.270      -0.414      1.552
               ----------------------------------------------------
```

CIs on the simple effect contrasts $W_1(b_1)$ and $W_1(b_2)$ (B6W1 and B7W1 in the output) show that both the new treatment and the standard treatment produce a substantial decrease in social anxiety between Pre and Post, and the CI on the subset effect contrast $W_1((b_1+b_2)/2)$ (B5W1 in the output) shows that the mean decrease for those who receive treatment (ignoring the type of treatment) is at least 0.92 standard deviation units. While there is evidence from the simple effect contrasts $W_2(b_1)$ (B6W2) and $W_2(b_2)$ (B7W2) of relapse between Post and FU2 following the termination of both treatments, there is good evidence of substantial relapse (at least 1.15 standard deviation units) only for the standard treatment. The between-subjects simple effect contrast $B_2(w_4)$ (B2W5 in the output) shows that the mean social anxiety score on the final measurement occasion is lower for the new treatment than for the standard treatment, but the estimate of the magnitude of difference is not particularly informative.

The range of contrasts discussed in the previous paragraph (simple effect contrasts of both types, interaction contrasts and a subset effect contrast) illustrates the flexibility of an analysis restricted only by the requirement that all contrasts must be $B \times W$ product contrasts. Apart from flexibility, the main feature of this single-family analysis is that variation in contrast standard errors is the only source of variation in precision of estimation, because the same CC is used to construct all CIs.

Mixed designs with more than two factors

Multiple between-subjects factors Data from mixed designs with more than one between-subjects factor can be analysed by applying the principles discussed in Chapter 5 to the between-subjects part of the analysis. Consider, for example, a $2 \times 2 \times (2)$ [$A \times B \times (C)$] design, with four groups (the product of the numbers outside the parentheses) and two measurements per subject. Assuming that values of the group-membership variable vary from 1 to 4 in the standard order $(a_1b_1, a_1b_2, a_2b_1, a_2b_2)$, a standard factorial analysis can be carried out with *PSY*, with the following contrasts in the input file:

```
[BetweenContrasts]
1  1 -1 -1 A
1 -1  1 -1 B
1 -1 -1  1 AB
[WithinContrasts]
1 -1 (C)
```

It would be necessary to carry out two *PSY* analyses. The first analysis (with all default options in place) would produce appropriate scaling for contrasts A, B, (C), A(C) and B(C), because the interaction between any correctly scaled Between contrast and any correctly scaled Within contrast will itself be correctly scaled by the program [as first-order interaction contrasts in the case of A(C) and B(C)]. The second analysis (with a scaling choice of *Interaction Contrasts: Between order 1*) would produce appropriate scaling for contrasts AB and AB(C). Output from this second analysis would include the following:

```
Individual 95% Confidence Intervals
-----------------------------------
The coefficients are rescaled if necessary
 to provide a metric appropriate for interaction contrasts.
Order of interaction for Between contrasts: 1
Order of interaction for Within contrasts:  0
```

Selection of an interaction scaling for Between contrasts does not change the scaling of Within contrasts.

Multiple within-subjects factors Data from mixed designs with more than one within-subjects factor can be analysed by applying the principles discussed in Chapter 6 to the within-subjects part of the analysis. Consider, for example, a $2 \times (2 \times 2)$ $[A \times (B \times C)]$ design, with two groups and four measurements per subject (the product of the numbers inside the parentheses). Assuming that the data section of the input file contains a group-membership variable (with values of 1 and 2) followed by four measurements (in the order $b_1c_1, b_1c_2, b_2c_1, b_2c_2$), a standard factorial analysis can be carried out with *PSY*, with the following contrasts in the input file:

```
[BetweenContrasts]
1 -1 A
[WithinContrasts]
1  1 -1 -1  (B)
1 -1  1 -1  (C)
1 -1 -1  1  (BC)
```

The first of two analyses (with all default options in place) would produce appropriate scaling for contrasts A, (B), (C), A(B) and A(C), leaving both interaction contrasts involving a within × within interaction component [namely (BC) and A(BC)] incorrectly scaled. The second analysis (with a scaling choice of *Interaction Contrasts: Within order 1*) would produce appropriate scaling for (BC) and A(BC) contrasts.

Complex mixed designs with multiple levels on some factors

The CCs required for *PSY* analyses controlling the FWER for main and interaction effect families of contrasts in standard unrestricted analyses can be calculated from general expressions given by Harris (1994, pp. 365-367). CCs are as follows:

Effect	*CC*	
Between-subjects	$\sqrt{\nu_B F_{\alpha;\nu_B,\nu_E}}$	(7.8)
Within-subjects	$\sqrt{\frac{\nu_W \nu_E}{\nu_E - \nu_W + 1} F_{\alpha;\nu_W,\nu_E-\nu_W+1}}$	(7.9)
Between × Within	$\sqrt{\frac{\nu_E\,\theta_{\alpha;s,m,n}}{1-\theta_{\alpha;s,m,n}}}$	(7.10)

where ν_B and ν_W are degrees-of-freedom parameters referring respectively to the between-subjects and the within-subjects component of the effect;

and *GCR* parameters are $s = \min(\nu_B, \nu_W)$, $m = (|\nu_B - \nu_W| - 1)/2$ and $n = (\nu_E - \nu_W - 1)/2$.

When $s = 1$, (7.10) is equivalent to (7.8) if $\nu_W = 1$ or to (7.9) if $\nu_B = 1$.

For example, a $2 \times 3 \times (4 \times 5)$ $[A \times B \times (C \times D)]$ experiment with $N = 60$ subjects has 6 groups, 20 observations per subject, and $\nu_E = N - 6 = 54$ degrees of freedom for error. For the $(C \times D)$ interaction involving the two within-subjects factors, ν_B is irrelevant and the CC is calculated from (7.9) with ν_W = (4 – 1)(5 – 1) = 12. For the second-order $B \times (C \times D)$ interaction involving one between-subjects factor and two within-subjects factors, the CC is obtained from (7.10) with *GCR* parameter values calculated from ν_B = (3 – 1) = 2 and ν_W = (4 – 1)(5 – 1) = 12. Question 2 in the *Questions and exercises* section deals with the CCs for all of the main and interaction effects in this analysis.

The general expressions for CCs given in this section [(7.8), (7,9) and (7.10)] can be used for the construction of SCIs in unrestricted CI analyses of data from any factorial design of any degree of complexity.

Beyond multifactor ANOVA

In this book we have dealt with CI analyses of (approximately) normally distributed data from single-factor and multifactor fixed-effects designs. While these analyses cover what many researchers might regard as the whole range of ANOVA-model analyses, they certainly do not cover the range of linear models (such as unsaturated models, models including covariates and multilevel models) available for the analysis of data from other types of designs. Nor do they deal properly with data for which the standard ANOVA-model assumptions do not (even approximately) apply. Some of the books recommended in earlier chapters provide introductions to a number of extensions of and alternatives to

ANOVA analyses, usually with an emphasis on significance tests rather than CIs. Cohen, Cohen, West and Aiken (2003) provide an excellent non-mathematical treatment of a wide range of linear-model analyses, with discussions of CI procedures as well as significance tests.

Further reading

O'Brien and Kaiser (1985) provide a detailed account of the MANOVA-model method of analysing data from mixed designs. For an examination of the issues involved in choosing a MANOVA test statistic when the FWER is to be controlled in unrestricted analyses of product contrasts, see Bird and Hadzi-Pavlovic (1983) or Harris (2001).

Questions and exercises

1. The social anxiety data set was produced by drawing 30 random samples from each of three 4-variate multivariate normal distributions with common covariance matrix

$$\Sigma = \begin{bmatrix} 40 & 25 & 28 & 25 \\ 25 & 45 & 38 & 36 \\ 28 & 38 & 50 & 44 \\ 25 & 36 & 44 & 60 \end{bmatrix}.$$

The population means matrix is

$$\mu = \begin{bmatrix} 35 & 25 & 27 & 29 \\ 35 & 25 & 30 & 35 \\ 35 & 35 & 35 & 35 \end{bmatrix}.$$

Given these means, the population values of the contrasts used in the first analysis of the social anxiety data set (Example 7.1) are as follows:

	W_0	W_1	W_2	W_3
B_0	–	$4.\dot{3}$	–3.5	$-2.\dot{3}$
B_1	–4.875	6.5	–5.25	–3.5
B_2	–2.25	3	4.5	3

(a) How many noncoverage errors (if any) are there in the set of raw CIs reported in Example 7.1? Comment on the implications of any noncoverage errors for the interpretation of the data.

(b) The population covariance matrix Σ shows that the within-population variances ($\sigma_1^2 = 40$, $\sigma_2^2 = 45$, $\sigma_3^2 = 50$ and $\sigma_4^2 = 60$) increase over time. The standard deviation required for the purpose of defining a standardized effect size is the square root of the mean of these variances, namely

$$\sigma = \sqrt{\frac{\sigma_1^2 + \sigma_2^2 + \sigma_3^2 + \sigma_4^2}{4}} = 6.982.$$

Given this value of σ, the population values of standardized contrasts (ψ_{gh}/σ) are as follows:

	W_0	W_1	W_2	W_3
B_0	–	0.621	–0.501	–0.334
B_1	–0.698	0.931	–0.752	–0.501
B_2	–0.322	0.430	0.645	0.430

How many noncoverage errors (if any) are there in the set of approximate standardized CIs reported in Example 7.1? Comment on the implications of any noncoverage errors for the interpretation of the data.

(c) It could be argued that the pre-treatment standard deviation would provide a more appropriate basis for standardization because it cannot be influenced by the treatments and it is necessarily the same for all three populations. Population values of contrasts standardized on the basis of pre-treatment variability $(\psi_{gh}/\sigma_{\text{Pre}})$ are as follows:

	W_0	W_1	W_2	W_3
B_0	–	0.685	–0.553	–0.369
B_1	–0.771	1.028	–0.830	–0.553
B_2	–0.356	0.474	0.712	0.474

What difference does the alternative basis for standardization make to the magnitudes of standardized contrast values in this particular case? Why does it make this difference?

2. A $2 \times 3 \times (4 \times 5)$ $[A \times B \times (C \times D)]$ balanced experiment with two between-subjects factors and two within-subjects factors has $N = 60$ subjects and $\nu_E = N - 6 = 54$ degrees of freedom for error. Calculate the CCs for unrestricted 95% SCIs within each of the main and interaction effects defined in a standard factorial analysis of the data.

Notes

1. The population *covariance* between ε_1 and ε_2 is the mean product of ε_1 and ε_2 values. The population covariance between two variables that do not have population means of zero is the mean product of deviation scores on those variables. The *correlation* between two variables is the covariance between standardized scores on those variables.

2. The ANOVA model that underlies the univariate approach to the analysis of repeated measures data assumes *compound symmetry*, implying that all of the variances in the covariance matrix Σ_Y are homogeneous and that all of the covariances in the same matrix are homogeneous. Compound symmetry is a sufficient but not necessary condition for *sphericity*, which implies that all normalized within-subjects contrasts have the same population variance. The MANOVA-model approach does not require either the very strong compound symmetry assumption or the slightly weaker (but still problematic) sphericity assumption.

3. Terms like post hoc and unrestricted refer to the definition of contrasts within families, not to the definition of the families themselves. An experimenter who wishes to inspect the data before deciding on whether to include B and/or W simple effect contrasts in an analysis should carry out a single-family analysis that allows for all factorial contrasts.

Appendix A *PSY*

Getting started

PSY runs under Windows (95 or later). It can be downloaded from

http://www.psy.unsw.edu.au/research/psy.htm

All of the downloaded files should be kept in the same directory. If you want to be able to open *PSY* from the desktop, right-click on the file *psy.exe*, select *Create Shortcut*, then move the shortcut to the desktop.

To start the program, click on the *PSY* logo (on the desktop) or on the file *psy.exe* (in the *PSY* directory). You should see the main *PSY* window including a menu bar (containing the commands *File*, *Edit*, *Calculate*, *View*, *Options*, *Window* and *Help*) and a toolbar including a number of standard Windows buttons, as well as two *PSY*-specific buttons:

Run Analysis, highlighted only when a *PSY Input File* has been created or opened;

Probability Calculator, a calculator that produces critical values and p values from central t, F, *GCR* and *SMR* distributions.

The Probability Calculator is not required for basic *PSY* analyses, but it is required for some advanced analyses.

The PSY input file

A *PSY* analysis requires an input file (**.in*), which can be constructed within the program (*File* → *New*) or from a text file (*File* → *Open* → **.txt*). A previously saved input file can be opened for subsequent analyses (*File* → *Open* → **.in*). To see a template for the input file, click on the *New File* button (the first button on the toolbar) or select *File* → *New* from the menu bar. You should then see an IN window, including [Data], [BetweenContrasts] and [WithinContrasts].

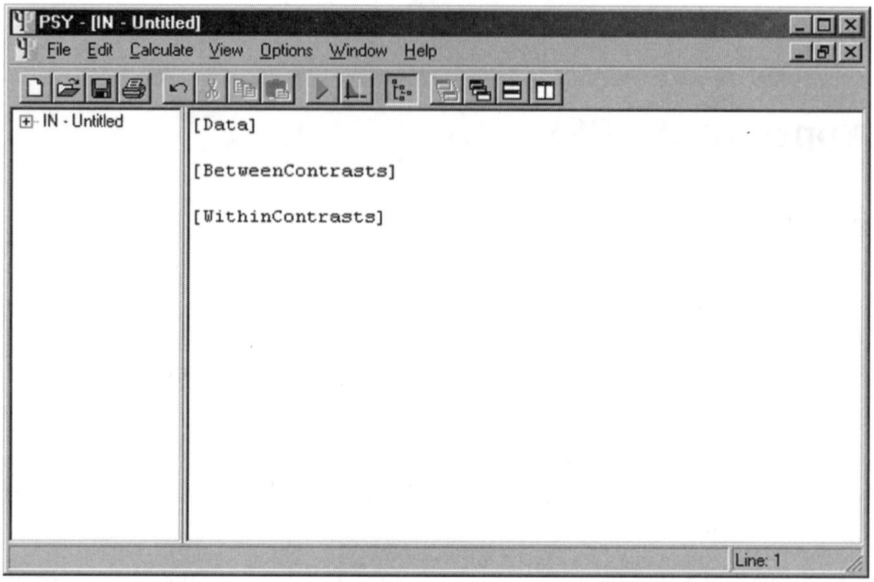

The input file must contain data under the heading [Data], and at least one set of contrast coefficients: a set of integer coefficients referring to groups (under the heading [BetweenContrasts] or [BContrasts]) and/or a set of integer coefficients referring to (repeated) measurements (under the heading [WithinContrasts] or [WContrasts]). If your analysis includes only contrasts of one type (B or W, but not both), you can either delete the irrelevant heading, or simply ignore it.

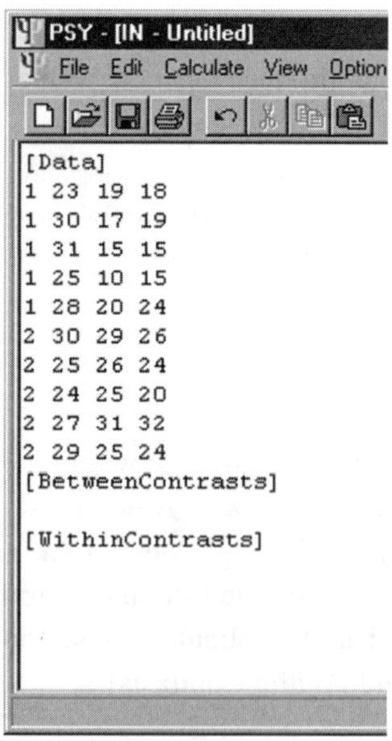

Data Data must be entered in free format with one row per subject, beginning with a subject in group 1, followed by the remaining subjects in that group. For every subject the first column of data is the value of a group membership variable specifying the number of the group (1, 2, 3, …) to which that subject belongs. If the experiment includes only one group, the value of this 'variable' (a constant in this case) must be 1. If the experiment includes two or more groups, enter the data for subjects in group 2 below the data for all subjects in group 1. Data for subjects in group j should be entered after the data for all subjects in earlier groups [1, 2, … , (j – 1)].

For every subject the second value (separated from the first by at least one space) must be that subject's first measurement (that is, the dependent variable score on the first measurement occasion). In a between-groups design without repeated measurements, this will be the only measurement for that subject. Otherwise, second and subsequent measurements are entered (in order) after the first, with at least one space between adjacent values. If data (but not contrasts) are entered from an experiment including two groups, five subjects per group and three measurements per subject, the left hand side of the IN window (after *View* → *Navigator* are selected) should look like that shown here.

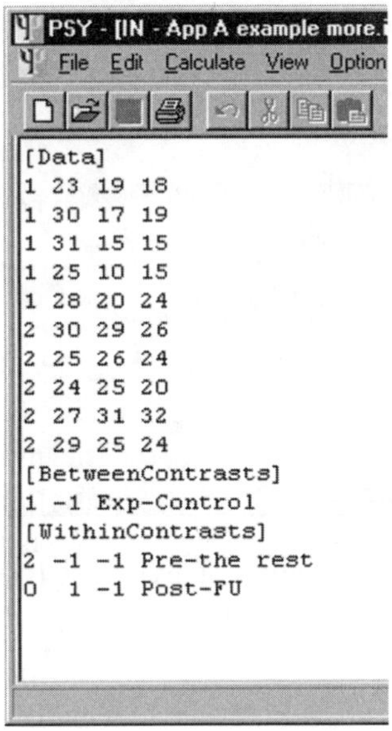

Contrasts At least one contrast must be defined in the input file. If the analysis includes one or more between-subjects contrasts, then each of those contrasts must appear in a new row under the [BetweenContrasts] heading, with one integer (whole number) coefficient for each group.

Adjacent coefficients must be separated by at least one space. A label (optional) of up to 12 characters can be entered after the space following the last coefficient of each contrast.

Similarly, if the analysis includes one or more within-subjects contrasts, then each of those contrasts must appear in a new row under the [WithinContrasts] heading, with one coefficient for each measurement. The input file shown here includes one between contrast and two within contrasts.

Examples of input files (or parts thereof) are given in Chapter 2 (Examples 2.1 and 2.3), Chapter 4 (Example 4.1), Chapter 5 (Examples 5.1 and 5.2), Chapter 6 (Examples 6.1 and 6.2) and Chapter 7 (Example 7.1).

Copying data from another application You can copy data within another application (such as *SPSS*) and paste it into a *PSY* input file. If you find that copied data uses a character (rather than a space) to mark the boundary between adjacent values, try running *PSY* before committing yourself to extensive editing. (*PSY* recognizes a number of alternative delimiters.) Make sure, however, that the order of subjects meets the requirements outlined above.

Running the analysis

When you have constructed the input file, save it as a **.in* file (optional) and click on the green *Run Analysis* button. You should then see the *Analysis Options* window.

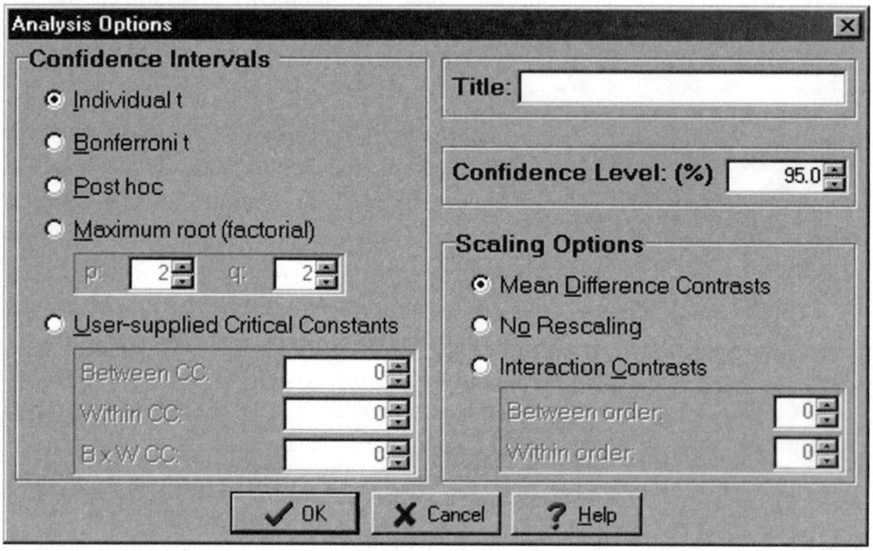

If you are happy with the default settings (which produce 95% individual CIs), type in a title for the analysis (optional) and click on the *OK* button. You should then see the output.

Output

The output file includes

- group (and measurement) means and standard deviations;
- an ANOVA-style summary table including an *F* ratio for each contrast;
- a table of CIs on raw contrast values;
- a table of approximate CIs on standardized contrast values.

The options chosen from the *Analysis Options* window have no influence on the ANOVA summary table; they determine the scaling applied to contrast coefficients (and therefore to the contrasts defined by those coefficients), the procedure used to construct CIs and the confidence level (and therefore error rates) associated with CIs. The output includes a summary of the implications of the chosen options for the CI tables. The output file from a default analysis based on the input file shown here includes the following:

```
                   Individual 95% Confidence Intervals
                   -----------------------------------
                   The CIs refer to mean difference contrasts,
                    with coefficients rescaled if necessary.
                    The rescaled contrast coefficients are:

                   Rescaled Between contrast coefficients
            Contrast        Group...
                             1            2
 Exp-Control  B1           1.000       -1.000

                   Rescaled Within contrast coefficients
            Contrast        Measurement...
                             1            2             3
 Pre-the rest W1           1.000       -0.500        -0.500
 Post-FU      W2           0.000        1.000        -1.000
```

In this case, coefficient rescaling has been necessary only for the first within-subjects contrast (W1).

Results in the output file are shown to 3 decimal places, unless you increase this number (by selecting *Options* → *Decimal Places* prior to running the analysis). The new setting will be used for future analyses until you change it again.

You can print and/or save the output file, or copy the contents and paste them into another application such as a *Word* file.

The Probability Calculator

The Probability Calculator can be selected from the menu bar (*Calculate* → *Probability Calculator*) or by clicking on the Probability Calculator button. The calculator is always available when *PSY* is open, whether or not you have opened an input file. Before using the calculator, you may wish to change the number of decimal places shown as the *Result* of the calculation (by selecting *Option* → *Decimal Places* from the menu bar).

The Probability Calculator provides unadjusted or Bonferroni-adjusted critical values or p values for central t, F, *GCR* and *SMR* distributions. You will not need to use the calculator to carry out *basic* CI analyses (those that can be implemented simply by selecting options from the Analysis Options window). If you wish to report p values for tests on contrasts (a practice not recommended in this book), however, you may need to use the Probability Calculator to calculate those values.

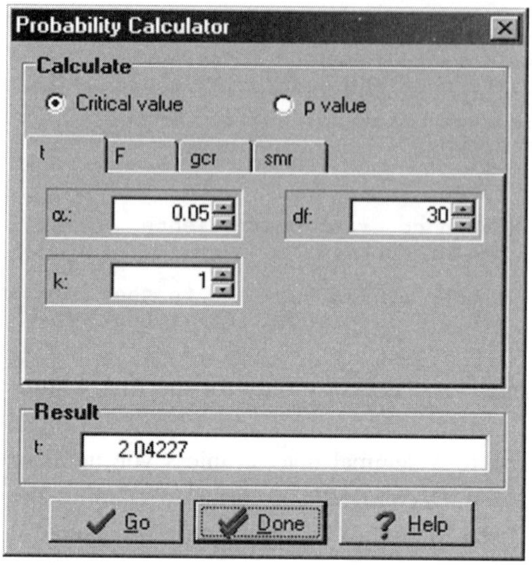

Critical value calculations

Some advanced *PSY* analyses require one or more *User-supplied Critical Constants*, which are functions of α-level critical values of *t*, *F*, *GCR* or *SMR* distributions. To obtain a .05-level critical value, select the distribution, enter the distribution parameter values [*df* (*t*), *df*1 and *df*2 (*F*), *s*, *m* and *n* (*GCR*), or *p*, *q* and *df* (*SMR*)] and click on *Go*. Otherwise, enter appropriate values for α and/or *k* (for a Bonferroni-adjusted α/*k*-level critical value) as well as distribution parameter values. Note that in the case of *t*-based CI procedures, the obtained critical value is also the required CC. In the case of *F*, *GCR* and *SMR* CI procedures, further calculations (which depend on the obtained critical value) are required to determine the appropriate CC.

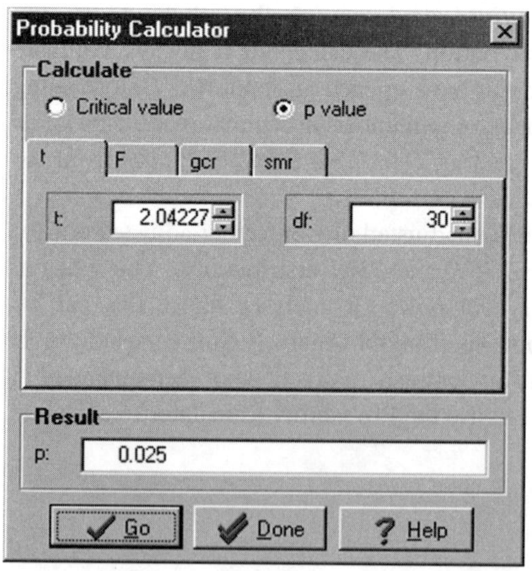

P value calculations

Select the distribution, click on *p value*, specify the distribution parameters and a value from the distribution and click on *Go*.

The resulting *p* value refers to the area in the distribution above the entered value. In the case of the *t* distribution (the only one of the four distributions where the α value is split between the two tails of the distribution for the purpose of determining a critical value for a two-sided CI or a two-tailed test), this means that the *p value* is a 'one-tailed' value, whereas the *critical value* is appropriate for a two-sided CI or a two-tailed test.

If you want to calculate the p value for a contrast from the F value (with $df1 = 1$) given in the ANOVA summary table, you should be aware that a comparison of the p value with an α level (such as .05) is directly relevant only for a planned test carried out with a PCER, and is completely compatible only with an *individual* (as distinct from simultaneous) CI.

Basic analyses

To carry out a basic *PSY* analysis, choose one of the first three options on the Analysis Options window (*Individual t*, *Bonferroni t* or *Post hoc*) and (if the default confidence level of 95% is acceptable) click on the *OK* button. A basic analysis is usually appropriate for the analysis of single-factor between-subjects or within-subjects designs, and for a standard analysis of a mixed design with one between-subjects factor and one within-subjects factor.

Single-factor between-subjects designs In a basic *PSY* analysis, 95% CIs are constructed on mean difference versions of the contrasts defined in the input file. To obtain individual CIs (the default option), click on the *OK* button. To obtain SCIs, select either the *Bonferroni t* or *Post hoc* option and then click on the *OK* button. The *Post hoc* option provides Scheffé intervals.

See Examples 2.1, 2.2 and 2.3.

Single-factor within-subjects designs When the Input file includes only within-subjects contrasts, a multivariate model is adopted for the analysis and each contrast has its own error term. If the *Post hoc* option is chosen, *PSY* produces MANOVA-model (Hotelling's T^2) SCIs.

Mixed designs with one between- and one within-subjects factor When the input file includes some *B* (between-subjects) and some *W* (within-subjects) contrasts, *PSY* will produce CIs on *B* main effect, *W* main effect and *BW* product interaction contrasts. The default scaling option (*Mean Difference Contrasts*) is applied to all main effect contrasts and the *BW* interaction contrasts are scaled so that they can be interpreted as a *B* mean difference in a *W* mean difference.

The *Bonferroni t* option controls the FWER in planned analyses for each of the three families of contrasts (*B*, *W* and *BW*). The *Post hoc* option controls the FWER for each family by constructing Scheffé SCIs on *B* main effect contrasts, Hotelling's T^2 SCIs on *W* main effect contrasts and Roy's *GCR* (MANOVA-model) SCIs on *BW* interaction contrasts.

See Examples 7.1 and 7.2.

Advanced analyses

Advanced *PSY* analyses require one or more of the remaining options in the *Analysis Options* window. If the user supplies the critical constant (CC), *PSY* can be instructed to construct central CIs of the form

$$\psi_g \in \hat{\psi}_g \pm \text{CC} \times \hat{\sigma}_{\hat{\psi}_g}$$

on any contrast defined on the parameters of a means model or a saturated ANOVA model.

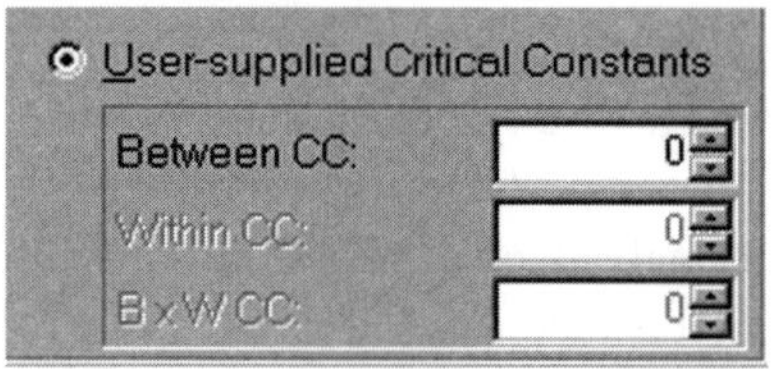

If you want to provide the program with a CC (or CCs), select *User-supplied Critical Constants*, and enter the CC (or CCs) in the highlighted field (or fields). If the input file contains only one type of contrast (Between or Within), then the corresponding CC will be highlighted. Otherwise, all three CCs (Between, Within and $B \times W$) will be highlighted. If all three are highlighted and the same CC is required for all contrasts in the analysis, then enter the required CC three times.

Examples 5.1, 6.2 and 7.3 illustrate the use the *User-supplied Critical Constants* option.

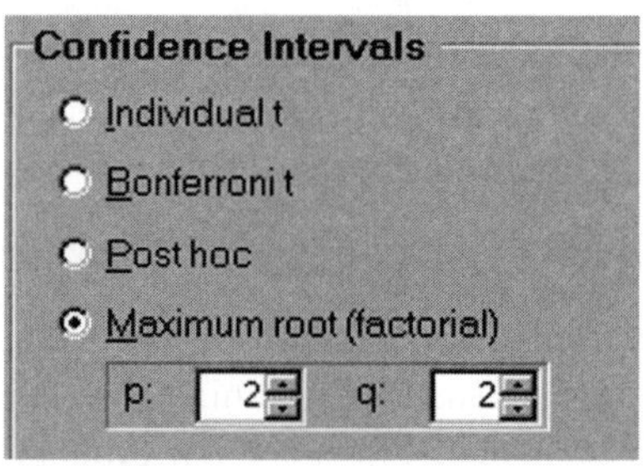

If a studentized maximum root (*SMR*) CC is required for an analysis, *PSY* can calculate the CC if the user supplies the p and q parameters of the required *SMR* distribution. Select *Maximum root (factorial)* and enter the p and q parameters.

Example 5.2 (Chapter 5) uses the *Maximum root (factorial)* option.

Scaling Options

The default scaling option (*Mean Difference Contrasts*) is appropriate for almost all contrasts (apart from interaction contrasts) defined on the parameters of ANOVA or means models. The most obvious exceptions are those that arise when a contrast is to be interpreted as a parameter defined by a different model, such as a regression coefficient defined by the linear regression model. In cases like this it is usually appropriate to select the *No Rescaling* option, then multiply or divide the resulting contrast statistics by whatever factor is required to transform the results into the appropriate metric.

The *No Rescaling* option is used in the last analysis in Appendix D (Trend Analysis).

First-order interaction contrasts *PSY* is based on a model that recognizes the distinction between a between-subjects factor and a within-subjects factor, but it does not recognize distinctions among multiple between-subjects or multiple within-subjects factors. When there is only one measurement, a *PSY* analysis is based on a between-subjects means model. Similarly, when there is only one group, a *PSY* analysis is based on a within-subjects means model. As a consequence, contrasts for two-factor between-subjects designs must be defined with coefficients referring to cell means rather than factor levels (see Table 4.1, Chapter 4). This is also true of contrasts for two-factor within-subjects designs. Further, the coefficient scaling required for $B \times B$ (between × between) interaction contrasts and for $W \times W$ (within × within) interaction contrasts is not the default (mean difference) scaling provided by *PSY*.

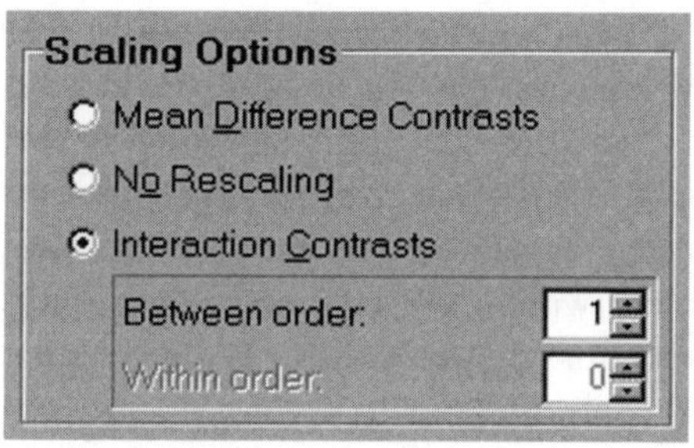

To apply a scaling appropriate for first-order contrasts to all between-subjects contrasts in a *PSY* analysis, select *Interaction Contrasts* from the *Scaling Options* on the Analysis Options window, then set the *Between order* (which will now be highlighted if there are any between-subjects contrasts in the Input file) to 1. If the input file also includes within-subjects contrasts, then both *Between order* and *Within order* will be highlighted. You can set both values independently.

Because *PSY* scales all contrasts of a particular type (between or within) in the same way in a given analysis, it is usually necessary to carry out two analyses to achieve the appropriate scaling for all contrasts in a two-factor between-subjects (or a two-factor within-subjects) design. If the two analyses differ only in scaling options, run the first analysis with the default scaling in place (thereby obtaining appropriately scaled main and/or simple effect contrasts), select interaction scaling for the second analysis, then edit the output file to retain the relevant CIs from both analyses. Alternatively, cut irrelevant contrasts from the input file before each run. In some analyses, other options (such as the CC for FWER control in a post hoc analysis) may need changing when scaling options are changed.

$B \times W$ interaction contrasts defined by the program are scaled correctly if the default option (*Mean Difference Contrasts*) is retained for the scaling of B and W contrast coefficient vectors (and main effect contrasts).

For examples of the use of the *Interaction Contrasts* scaling option, see Examples 4.1, 5.1, 5.2, 6.1 and 6.2.

Higher-order interaction contrasts involving factors of the same type Second-order (triple) interaction contrasts involving factors of the same type ($B \times B \times B$ or $W \times W \times W$) are scaled appropriately by selecting *Interaction Contrasts, Between order* = 2 or *Interaction Contrasts, Within order* = 2. In general, a higher-order interaction contrast involving factors of the same type is scaled appropriately if the order of interaction selected is one less than the number of factors involved in the interaction.

Higher-order interaction contrasts involving factors of both types Mixed designs with more than two factors can produce higher-order interactions involving multiple between-subjects and/or within-subjects factors. For example, second-order (triple) interaction contrasts in a $2 \times (3 \times 4)$ design involve one between-subjects and two within-subjects factors. To obtain the correct scaling of these contrasts in a *PSY* analysis, the user must ensure that all user-defined coefficient vectors (that is, all coefficient vectors appearing in the input file) are scaled appropriately. In the $2 \times (3 \times 4)$ case, the B coefficient vector [1 –1] is scaled correctly by the default option (mean difference scaling). Default scaling of the 12 element within-subjects coefficient vector would be not be appropriate, however, for first-order $W \times W$ interaction contrasts or for second-order $B \times (W \times W)$ contrasts.

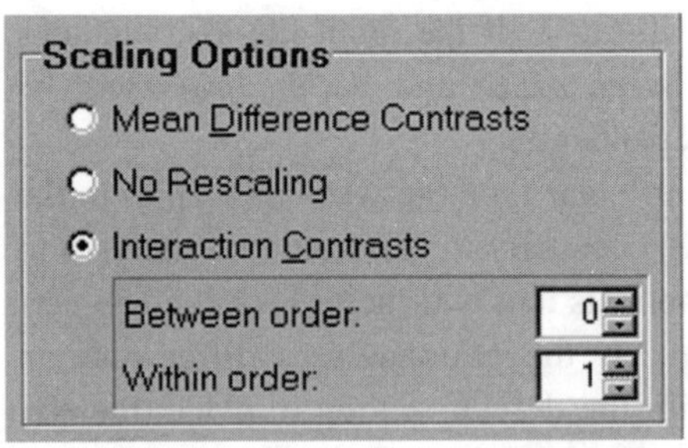

Selection of the *Interaction Contrasts* option will highlight both *Between order* and *Within order* options. Correct scaling for $B \times (W \times W)$ contrasts is achieved by accepting the *Between order* default of zero and setting the *Within order* to 1.

Similar principles apply to higher-order interaction contrasts of any degree of complexity. For example, a fourth-order interaction contrast involving all five factors in a $2 \times 2 \times 3 \times (4 \times 5)$ design would require *Between order* = 2 (to deal with the $2 \times 2 \times 3$ component) and *Within order* = 1 [to deal with the (4×5) component]. If the user specifies the correct order for both components of a mixed interaction in a mixed design, then *PSY* will scale the interaction appropriately.

Controlling FWERs in factorial designs With some exceptions, standard factorial analyses require multiple CCs (one for each family). For *planned* analyses using the Bonferroni-*t* procedure, the easiest way to carry out multiple *PSY* analyses is to cut irrelevant contrasts from the input file before each run and select *Bonferroni t* from the Analysis Options window. (This selection will be retained for all runs unless it is changed during the *PSY* session.) For *post hoc* analyses with more than one between-subjects (or more than one within-subjects) factor, it will often be necessary to provide *PSY* with the CC required for each run, because the appropriate CC will be a function of the number of degrees of freedom for the effect, rather than the number of contrasts in the analysis.

Multiple Bonferroni-*t* analyses are carried out in Example 6.1. Examples 5.1 and 6.2 illustrate the use of multiple user-supplied CCs in post hoc factorial analyses.

Nonstandard confidence levels

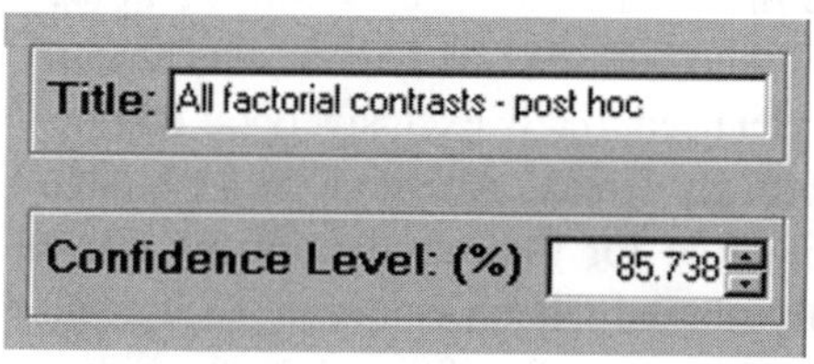

Factorial analyses allowing for direct inferences on simple and subset effect contrasts can sometimes require nonstandard confidence levels such as $100(1 - 3\alpha)$ or $100(1 - \alpha)^3$. To carry out these analyses, select *Confidence Level* and change the level to the required value.

Examples 4.1, 5.2 and 7.3 use nonstandard confidence levels.

Appendix B *SPSS*

While most statistical packages provide tests of homogeneity of the effect parameters in ANOVA models, very few provide SCIs on contrasts on those parameters. *SPSS* is a notable exception, with two GLM programs (*SPSS MANOVA* and *SPSS GLM*), one or both of which can provide individual or simultaneous raw CIs on contrasts defined on parameters of saturated or unsaturated ANOVA models. In some cases (specifically analyses based on between-subjects multifactor ANOVA models) it is easier to carry out CI analyses with *SPSS MANOVA* than with *PSY*. In other cases (specifically analyses based on within-subjects or mixed models), it is easier to use *PSY*.

SPSS analyses do not construct standardized CIs, nor do they construct *SMR*-based CIs. Nevertheless, if you have access to both *PSY* and *SPSS* you can carry out a wider range of CI analyses than if you need to rely exclusively on *PSY*.

SPSS MANOVA is accessible only through syntax. Although *SPSS GLM* is accessible through menus, syntax is required for the construction of CIs on contrasts. *SPSS GLM* will construct only *t*-based CIs on contrasts, so it cannot be used for post hoc analyses, or, more generally, for coherent analyses including standard homogeneity tests. For these reasons we will be concerned here primarily (but not exclusively) with *SPSS MANOVA*.

Between-subjects factorial designs

The *SPSS* data file must include one variable for each factor (*A*, *B*, *C*, and so on) and *Y*, the dependent variable. The values of each factor variable are levels (1, 2 or 3 for a three-level factor). A two-factor ANOVA-model post hoc analysis of the 3 × 4 (Fee × Treatment) data set discussed in Chapter 5 requires a data file with 192 cases and three variables: *A* (with values varying from 1 to 3), *B* (with values varying from 1 to 4) and *Y*. The following commands will produce homogeneity tests and the same raw Scheffé SCIs as those produced by the *PSY* analysis in Example 5.1.

```
manova y by A(1 3) B(1 4)
  /contrast(A)=special(1  1  1
                      .5 .5 -1
                       1 -1  0)
```

```
/contrast(B)=special(1   1   1   1
                     .5  .5 -.5 -.5
                     1  -1   0   0
                     0   0   1  -1)
/cinterval joint(.95) univariate(scheffe)
/print param(estim).
```

The *A* and *B* coefficient matrices must be square (that is, they must have the same number of rows as columns), the first row must contain uniform coefficients and the contrasts must be linearly independent. If the *A* and *B* coefficient vectors are scaled to define mean difference contrasts (as they are here), then the interaction contrasts in the output will be appropriately scaled.

Replacing the second last line of syntax with

```
/cinterval (.95) univariate
```

will produce individual CIs. The line

```
/cinterval joint(.95) univariate(bon)
```

will produce Bonferroni-*t* CIs.

SPSS does not provide analyses controlling the *FWER* for families such as *A*(*B*) defined by simple effect models.

Fitting an unsaturated model *SPSS MANOVA* produces an unsaturated-model analysis if the effects in that model are specified on a *design* line in the syntax file. To construct CIs on contrasts defined on parameters of the main effects model (4.2), remove the stop from the end of the print line and add the line

```
/design A B.
```

Within-subjects designs

The CI analyses discussed in Chapters 6 and 7 are more difficult to carry out with *SPSS* than with *PSY*, because the Repeated Measures options in *SPSS MANOVA* will not construct CIs on mean difference within-subjects contrasts, and *SPSS GLM* will not construct post hoc CIs.[1] It is possible, however, to persuade *SPSS MANOVA* to construct CIs on any variables, including contrast variables. Consider the small data set ($n = 5$, $p = 3$) used to illustrate single-factor within-subjects analyses in Chapter 7. If an *SPSS* data file contains scores on Y_1, Y_2 and Y_3, the following syntax produces 95% Bonferroni-*t* SCIs on all comparisons.

```
compute w1 = y2-y1.
compute w2 = y3-y2.
compute w3 = y3-y1.
manova w1 w2 w3
  /cinterval (.98333333)
  /print param(estim)
  /analysis w1 w2/design
  /analysis w3.
```

The *compute* commands define the contrast variables on which the analysis is to be based. Because *SPSS MANOVA* will accept only linearly independent measures, it is not possible to include all three contrast variables (which are not linearly independent) in a single analysis. The first analysis includes the maximum of $(p-1) = 2$ linearly independent contrast variables, while the second analysis includes the final contrast variable. The confidence level of $(1 - \alpha/k) = .98\dot{3}$ is provided on the fifth line of syntax in order to control the PFER at $\alpha = .05$, thereby ensuring that FWER < .05.

The Bonferroni command available for *SPSS MANOVA* cannot be used here because the three 'planned' (but linearly dependent) contrasts are not included in a single analysis. If Bonferroni-*t* intervals are to be constructed on three linearly independent contrast variables (which is possible only if $p \geq 4$), then Bonferroni-*t* intervals can be produced by following the appropriate set of *compute* commands (defining linearly independent contrast variables *w1*, *w2* and *w3*) with

```
/cinterval joint(.95) univariate(bon)
/print param(estim)
/analysis w1 w2 w3.
```

Replacing the *cinterval* line with

```
/cinterval joint(.95) multivariate
```

will produce an unrestricted (post hoc) analysis, with MANOVA (T^2) CIs.

Within-subjects factorial designs

Planned analyses Raw *t*-based CIs can be obtained from *SPSS MANOVA* or from *SPSS GLM*. The syntax required for an *SPSS MANOVA* analysis is cumbersome, however, if the contrasts are not linearly independent. *SPSS GLM* does not require contrasts to be linearly independent.

The planned analysis of the (3 × 4) Warnings study reported in Example 6.1 requires an *SPSS* data file including variables Y_{11}, Y_{12}, ..., Y_{34}. The following syntax produces the same raw Bonferroni-*t* SCIs as those produced by *PSY*.

```
GLM Y11 Y12 Y13 Y14 Y21 Y22 Y23 Y24 Y31 Y32 Y33 Y34
  /mmatrix = all 1/4 1/4 1/4 1/4 -1/4 -1/4 -1/4 -1/4 0 0 0 0;
             all 1/4 1/4 1/4 1/4 0 0 0 0 -1/4 -1/4 -1/4 -1/4;
             all 0 0 0 0 1/4 1/4 1/4 1/4 -1/4 -1/4 -1/4 -1/4;
  /criteria=alpha(.01666667).
GLM Y11 Y12 Y13 Y14 Y21 Y22 Y23 Y24 Y31 Y32 Y33 Y34
  /mmatrix = all 1/3 -1/3 0 0 1/3 -1/3 0 0 1/3 -1/3 0 0;
             all 1/3 0 -1/3 0 1/3 0 -1/3 0 1/3 0 -1/3 0;
             all 1/3 0 0 -1/3 1/3 0 0 -1/3 1/3 0 0 -1/3;
             all 0 1/3 -1/3 0 0 1/3 -1/3 0 0 1/3 -1/3 0;
             all 0 1/3 0 -1/3 0 1/3 0 -1/3 0 1/3 0 -1/3;
             all 0 0 1/3 -1/3 0 0 1/3 -1/3 0 0 1/3 -1/3;
  /criteria=alpha(.008333333).
```

```
GLM Y11 Y12 Y13 Y14 Y21 Y22 Y23 Y24 Y31 Y32 Y33 Y34
   /mmatrix = all 1 -1  0  0 -1  1  0  0  0  0  0  0;
              all 1  0 -1  0 -1  0  1  0  0  0  0  0;
              all 1  0  0 -1 -1  0  0  1  0  0  0  0;
              all 0  1 -1  0  0 -1  1  0  0  0  0  0;
              all 0  1  0 -1  0 -1  0  1  0  0  0  0;
              all 0  0  1 -1  0  0 -1  1  0  0  0  0;
              all 1 -1  0  0  0  0  0  0 -1  1  0  0;
              all 1  0 -1  0  0  0  0  0 -1  0  1  0;
              all 1  0  0 -1  0  0  0  0 -1  0  0  1;
              all 0  1 -1  0  0  0  0  0  0 -1  1  0;
              all 0  1  0 -1  0  0  0  0  0 -1  0  1;
              all 0  0  1 -1  0  0  0  0  0  0 -1  1;
              all 0  0  0  0  1 -1  0  0 -1  1  0  0;
              all 0  0  0  0  1  0 -1  0 -1  0  1  0;
              all 0  0  0  0  1  0  0 -1 -1  0  0  1;
              all 0  0  0  0  0  1 -1  0  0 -1  1  0;
              all 0  0  0  0  0  1  0 -1  0 -1  0  1;
              all 0  0  0  0  0  0  1 -1  0  0 -1  1;
   /criteria=alpha(.002777777).
```

Each of the three analyses defines within-subjects contrasts with a coefficient matrix (called an M matrix) referring to cell means. *SPSS GLM* does not rescale coefficient vectors, so the vectors of the M matrices must be scaled appropriately. That is, the sum of positive coefficients in each row of the main effect M matrices must be 1.0 (in order to define mean difference contrasts), and the sum of positive coefficients in each row of the third (interaction) M matrix must be 2.0 (so that each contrast can be interpreted as a difference in a difference). Fractional coefficients (as distinct from decimal coefficients) are acceptable. (It is easier and more accurate to enter values of 1/3 than it is to enter values of 0.333333.)

Nonstandard alpha values on the *criteria* lines are values of α / k_{Fam}, where k_{Fam} is the number of contrasts in the relevant family. These values ensure that the $100(1 - \alpha / k_{\text{Fam}})\%$ individual CIs constructed by *SPSS GLM* are also Bonferroni-*t* $100(1 - \alpha)\%$ SCIs.

The relevant section of the *SPSS* output file for each analysis (not shown here) appears under the major heading *Custom Hypothesis Tests*. The CIs appear under the secondary heading *Contrast Results* (*K matrix*) as intervals on transformed variables labelled T1, T2, and so on.

This *SPSS GLM* analysis provides no obvious advantages over the *PSY* analysis. The main disadvantage of the *SPSS GLM* analysis is that it does not provide standardized CIs.

Post hoc analyses *SPSS GLM* cannot produce T^2 SCIs (or tests) on within-subjects contrasts. It follows that this program cannot produce coherent analyses beginning with multivariate-model homogeneity tests, which it does produce. *SPSS MANOVA* can produce T^2 SCIs on within-subjects contrasts. The syntax required for unrestricted analyses within the standard *A*, *B* and *AB* families is straightforward provided that the analysis of each effect is based on exactly

ν_{Fam} linearly independent contrast variables. If this is not the case, then it will be necessary to carry out as many different (but possibly overlapping) analyses (with ν_{Fam} linearly independent contrast variables per analysis) as are required to ensure that every contrast is included in at least one analysis. The following syntax produces the same raw (but not standardized) T^2 CIs as those produced by *PSY*.

```
compute A1=(y11+y12+y13+y14)/4-(y21+y22+y23+y24+y31+y32+y33+y34)/8.
compute A2=(y21+y22+y23+y24)/4-(y31+y32+y33+y34)/4.
compute B1=(y11+y21+y31)/3-(y12+y22+y32)/3.
compute B2=(y11+y21+y31)/3-(y13+y14+y23+y24++y33+y34)/6.
compute B3=(y12+y22+y32)/3-(y13+y14+y23+y24++y33+y34)/6.
compute B4=(y13+y23+y33)/3-(y14+y24+y34)/3.
compute A1B1=(2*y11-y21-y31)/2-(2*y12-y22-y32)/2.
compute A1B2=(2*y11-y13-y14)/2-(2*y21-y23-y24)/4-(2*y31-y33-y34)/4.
compute A1B3=(2*y12-y13-y14)/2-(2*y22-y23-y24)/4-(2*y32-y33-y34)/4.
compute A1B4=(y13-y14)-(y23-y24)/2-(y33-y34)/2.
compute A2B1=(y21-y22)-(y31-y32).
compute A2B2=(2*y21-y23-y24)/2-(2*y31-y33-y34)/2.
compute A2B3=(2*y22-y23-y24)/2-(2*y32-y33-y34)/2.
compute A2B4=(y23-y24)-(y33-y34).
manova A1 A2 B1 B2 B3 B4 A1B1 A1B2 A1B3 A1B4 A2B1 A2B2 A2B3 A2B4
  /cinterval joint(.95) multivariate
  /print param(estim)
  /analysis A1 A2/design
  /analysis B1 B2 B4/design
  /analysis B1 B3 B4/design
  /analysis A1B1 A1B2 A1B4 A2B1 A2B2 A2B4/design
  /analysis A1B1 A1B3 A1B4 A2B1 A2B3 A2B4.
```

There is only one *A* main effect analysis because the $k_A = \nu_A = 2$ contrasts account for all of the variation within the *A* main effect without redundancy. There are two *B* analyses because there are more *B* main effect contrasts (4) than there are degrees of freedom for the effect (3). Each of the two analyses (the first excluding the contrast B_3, the second excluding B_2) includes $\nu_B = 3$ linearly independent contrasts, as it must to produce the correct T^2 CIs. If all four *B* contrasts were included in one analysis, the program would produce the following warning rather than CIs:

```
*  W A R N I N G  * The WITHIN CELLS error matrix is SINGULAR.  *
*                 * These variables are LINEARLY DEPENDENT       *
*                 * on preceding ones ..                         *
*                 *   B3                                         *
*                 * Multivariate tests will be skipped.          *
```

This message shows that B_3 must be dropped from the analysis if B_1 and B_2 are included.

Of the two analyses required to produce T^2 SCIs on all eight *AB* interaction contrasts, the first excludes the two contrasts involving B_3 and the second excludes the two contrasts involving B_2. These analyses would also tell us that the hypothesis of homogeneity of interaction means cannot be rejected by a .05-level multivariate T^2 test (p = .066), from which it follows that there can be no statistically significant difference on *any* interaction contrast in a coherent T^2 analysis.

This is a tedious analysis to implement, and it is not difficult to make errors when constructing the *compute* commands

Noncentral interval estimation

The noncentral interval estimation procedures described in Appendix C can be implemented by *SPSS* 11.0 (or later) with syntax files *NoncT2.sps* and *NoncF3.sps* provided by Michael Smithson at his website (http://www.anu.edu.au/psychology/staff/mike/CIstuff/CI.html).

To construct t-based noncentral standardized CIs on a set of planned contrasts, create an *SPSS* data file with variable names *tval*, *df* and *conf*. For each contrast (one contrast per row), enter the contrast t statistic, degrees of freedom for the t statistic, and the per-contrast confidence level (expressed as a probability) as values of these variables. Running the syntax file *NoncT2.sps* will produce additional variables in the data file, two of which (*lc2* and *uc2*) are the lower and upper limits of the CI on the noncentrality parameter for each contrast. Transform these limits into limits on standardized contrast values using the relevant expressions in Appendix C. For Bonferroni-t noncentral CIs on a set of k planned contrasts, set the value of *conf* at $(1 - \alpha/k)$.

To construct an F-based noncentral CI on a monotonic function of the noncentrality parameter δ, create an *SPSS* data file with variable names *fval*, *df1*, *df2* and *conf*. Enter the ANOVA F statistic, degrees-of-freedom parameters and confidence level (expressed as a probability) as values of these variables. Running the syntax file *NoncF2.sps* will produce additional variables in the data file, two of which (*lc2* and *uc2*) give the lower and upper limits of the CI on δ. Two other variables (*lr2* and *ur2*) give the corresponding limits on ω^2. CI limits on other monotonic functions of δ can be calculated by hand from the relevant expressions in Appendix C.

Note

1. The *Repeated Measures* option in *SPSS MANOVA* constructs CIs only on *orthonormalized* within-subjects contrasts. This means that if the user specifies an orthogonal set of $(p - 1)$ contrasts, the program will construct CIs on generally uninterpretable normalized contrasts (rescaled by setting $\sum c^2$ equal to 1.0). If $(p - 1)$ nonorthogonal contrasts are specified, the program will construct CIs on a similar (but not identical) set of orthogonal normalized contrasts. *SPSS GLM* does not use multivariate CCs to construct CIs, so it cannot construct intervals appropriate for post hoc analyses.

Appendix C Noncentral Confidence Intervals

The noncentral CIs discussed in this book refer to *standardized* parameters, not to raw parameters scaled in dependent variable units. Central CIs on raw parameters are exact (given the standard ANOVA-model assumptions). All of the following discussion refers to CIs on standardized parameters.

CIs based on noncentral F distributions

Given the ANOVA model $Y_{ij} = \mu + \alpha_j + \varepsilon_{ij}$ and the standard assumptions, the ANOVA F statistic from an experiment with J groups and equal ns has a noncentral F distribution $(F_{J-1,\, N-J,\, \delta})$ with noncentrality parameter

$$\delta = \frac{n\sum_j \alpha_j^2}{\sigma_\varepsilon^2}.$$

The lower limit of a 100(1 – α)% CI on δ is the noncentrality parameter δ_L of the distribution $F_{J-1,\, N-J,\, \delta_L}$ whose upper 100(1 – α/2)th percentile point is the obtained F statistic. The upper limit is the noncentrality parameter δ_U of the distribution $F_{J-1,\, N-J,\, \delta_U}$ whose upper 100(α/2)th percentile point is the obtained F statistic. A computer-intensive search procedure is required to find the noncentral F distributions $F_{J-1,\, N-J,\, \delta_L}$ and $F_{J-1,\, N-J,\, \delta_U}$.

Consider the Depression data set discussed in Chapter 2, with an obtained F statistic of $F_{3,76} = 5.91357$. The noncentral F distribution $F_{3,\, 76,\, 4.48317}$ (shown on the left of Figure C1) has 5% of its area above the obtained F value, so δ_L = 4.48317 is the lower limit of the 90% noncentral CI on δ. We can infer from this limit that $\delta \geq 4.48317$. The distribution $F_{3,\, 76,\, 32.90493}$ (shown on the right of Figure C1) has 95% of its area above the obtained F value, so $F_{J-1,\, N-J,\, \delta_U}$ = 32.90493 is the lower limit of the 90% noncentral CI on δ. We can infer from this limit that $\delta \leq 32.90493$. The two-sided CI (4.48317, 32.90493) is a 90% CI because each limit is the limit of a 95% single-sided CI.

The limits of the CI on δ can be transformed into limits of a CI on any monotonic function of δ, including the effect size parameters

$$f = \frac{\sigma_\alpha}{\sigma_\varepsilon} = \sqrt{\frac{\delta}{N}}$$

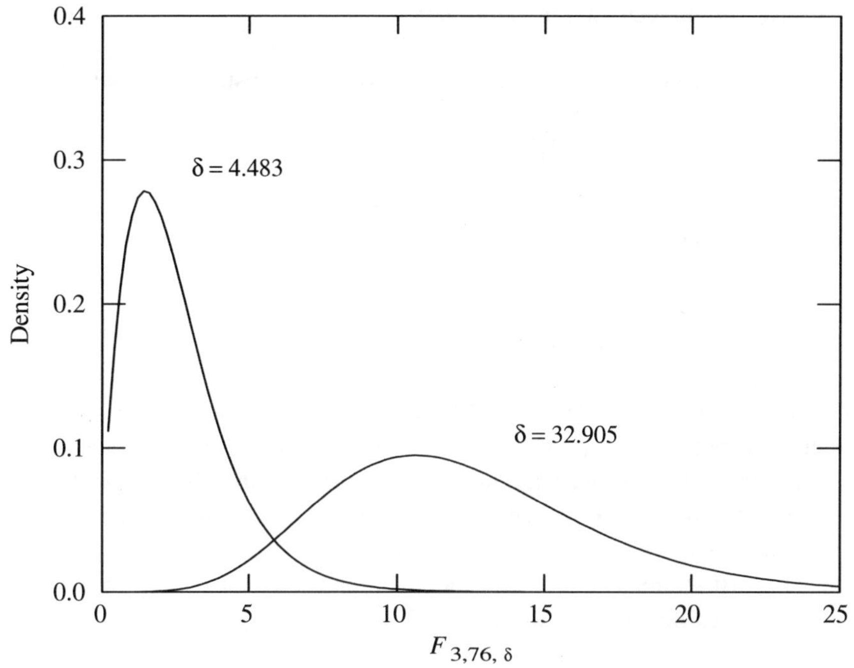

Figure C1 *Noncentral F distributions* $F_{3,\,76,\,4.483}$ *and* $F_{3,\,76,\,32.905}$

and $$\varpi^2 = \frac{\sigma_\alpha^2}{\sigma_\alpha^2 + \sigma_\varepsilon^2} = \frac{f^2}{1+f^2} = \frac{\delta}{N+\delta}.$$

In a model-comparison application where the unsaturated no-effects model $Y_{ij} = \mu + \varepsilon_{ij}$ is to be compared with the unsaturated ANOVA model, the CI limits on δ_F can be transformed into McDonald's (1997) goodness-of-approximation index

$$\Lambda = \sqrt{\frac{N}{N+\delta}}\,.$$

Values of δ_L and δ_U can be obtained from *STATISTICA Power Analysis* (by selecting *Interval Estimation / Several Means / ANOVA / 1-Way* or, if *n*s are not equal, by selecting *Probability Distributions / Noncentral F Distribution*).

For details of an *SPSS* implementation, see Appendix B.

Problems with noncentral CIs on δ If the obtained *F* value is smaller than $F_{\alpha/2;\,J-1,N-J}$, then δ_L is undefined. If the obtained *F* value is smaller than $F_{1-\alpha/2;\,J-1,N-J}$, then both δ_L and δ_U are undefined. These problems arise from the fact that δ is a function of squared effect parameters and therefore cannot be negative.

CIs based on noncentral *t* distributions

Let ψ_g be a planned between-subjects contrast. The statistic

$$t_{\psi_g} = \hat{\psi}_g \Bigg/ \sqrt{MSE \sum_j \frac{c_{gj}^2}{n_j}}$$

has a noncentral *t* distribution (t_{N-J,δ_g}) with noncentrality parameter

$$\delta_g = \frac{\psi_g}{\sigma_\varepsilon} \Bigg/ \sqrt{\sum_j \frac{c_{gj}^2}{n_j}}.$$

The lower limit of a 100(1 – α)% CI on δ_g is the noncentrality parameter δ_L of the distribution t_{N-J,δ_L} whose upper 100(1 – α/2)th percentile point is t_{ψ_g}. The upper limit is the noncentrality parameter δ_U of the distribution t_{N-J,δ_U} whose upper 100(α/2)th percentile point is t_{ψ_g}. A computer-intensive search procedure is required to find the noncentral *t* distributions t_{N-J,δ_L} and t_{N-J,δ_U}. Once the distributions have been found, the limits of the CI $\delta_g \in (\delta_L, \delta_U)$ can be transformed into limits *ll* and *ul* of a CI on $\psi_g/\sigma_\varepsilon$, by calculating

$$ll = \delta_L \sqrt{\sum_j \frac{c_{gj}^2}{n_j}} \quad \text{and} \quad ul = \delta_U \sqrt{\sum_j \frac{c_{gj}^2}{n_j}}.$$

For equal *n*s, this procedure is implemented by *STATISTICA Power Analysis* (*Interval Estimation / Several means / Planned Contrast*) which provides CIs on δ_g and $\psi_g/\sigma_\varepsilon$, given *J*, *n*, $\sum c_{gj}^2$ and t_{ψ_g}. For unequal *n*s, δ_L and δ_U can be obtained from *Probability Distributions / Noncentral t Distribution*.

For details of an *SPSS* implementation, see Appendix B.

Comparing noncentral and central t-based CIs on standardized contrasts

Exact noncentral *t*-based CIs on standardized planned contrasts should be preferred to approximate central *t*-based CIs when both are available. There are no noncentral SCI procedures for post hoc analyses, however, so post hoc procedures based on central *F*, T^2, *SMR* and *GCR* distributions have no noncentral competitors.

Comparisons of central and noncentral *t*-based standardized CIs (Table C1) show that the two approaches produce very similar intervals when the midpoint of the central interval is less than about 1.0 (a 'large' estimated value of Cohen's *d*). When the estimated effect is larger than about 1.0, central CIs are noticeably too narrow (and therefore the noncoverage error rate must be greater than α), particularly when the sample size is relatively small.

Table C1 *Central and noncentral t-based CIs on* $(\mu_1 - \mu_2)/\sigma_\varepsilon$ *when* $J = 2$

(a) $n = 10$

$\frac{M_1 - M_2}{\hat{\sigma}_\varepsilon}$	Approximate central CI		Exact noncentral CI	
	Lower limit	Upper limit	Lower limit	Upper limit
0	–0.94	0.94	–0.88	0.88
0.5	–0.44	1.44	–0.40	1.39
1.0	0.06	1.94	0.05	1.92
1.5	0.56	2.44	0.48	2.49
2.0	1.06	2.94	0.89	3.07
2.5	1.56	3.44	1.29	3.68

(b) $n = 50$

$\frac{M_1 - M_2}{\hat{\sigma}_\varepsilon}$	Approximate central CI		Exact noncentral CI	
	Lower limit	Upper limit	Lower limit	Upper limit
0	–0.40	0.40	–0.39	0.39
0.5	0.10	0.90	0.10	0.90
1.0	0.60	1.40	0.58	1.41
1.5	1.10	1.90	1.05	1.94
2.0	1.60	2.40	1.52	2.48
2.5	2.10	2.90	1.97	3.02

Relationships between F-based and t-based noncentral CIs When $J = 2$ (and, more generally, when an ANOVA effect has 1 degree of freedom), the square of the obtained t statistic is an F statistic with $\nu_1 = 1$, and the square of a t noncentrality parameter (such as δ_g) is the corresponding F noncentrality parameter. It might seem reasonable, then, to expect that the F-based noncentral CI would in some sense correspond to the t-based noncentral CI when $\nu_1 = 1$. This is not always the case. A t noncentrality parameter can be negative and has no lower bound. Similarly, a noncentral CI on δ_g has one negative limit (and one positive limit) whenever the corresponding F-based CI has no defined lower limit. (In practice, the lower limit of an F-based noncentral CI is set at zero when in theory it is not defined.)

The relationship between a noncentral CI based on a t statistic and that based on the corresponding $F = t^2$ statistic is reasonably straightforward when almost all of the values in both of the relevant noncentral t distributions have the same sign. In that case, each of the CI limits on the F noncentrality parameter is approximately equal to the square of the corresponding limit on the t noncentrality parameter.

Appendix D Trend Analysis

In some randomized experiments the J distinct treatments represent different values on a single quantitative dimension. Suppose, for example, that the treatments in a four-group experiment are different levels of sleep deprivation (X = 0, 12, 24 or 36 hours). After being deprived of sleep for the relevant number of hours, the performance of each subject on a simulated driving task (Y) is assessed. The experimenter is likely to be interested in asking about the form and magnitude of the relationship between X and Y. If the relationship appears to be *linear*, so that an increase of one hour in the level of sleep deprivation has the same effect on Y at all levels of X within the range examined in the experiment, then the experimenter would be interested in estimating the magnitude of that linear effect. This is the kind of effect defined by a *linear regression* model rather than an ANOVA model.

Linear regression analysis

The linear regression model is

$$Y = \beta_0 + \beta_1 X + \varepsilon \qquad \text{(D1)}$$

where the *regression coefficient* (β_1) can be interpreted as the size of the effect on Y of a unit increase (an increase of 1.0) in X. When $J > 2$, the linear regression model is an unsaturated model that is unlikely to fit the population means perfectly.

The text file *sleep.txt* contains artificial data from the sleep deprivation experiment described above with ten subjects per group ($J = 4$, $n = 10$, $N = 40$). The two variables in the file are X (with values of 0, 12, 24 or 36) and an error score Y. Sample means are $M_0 = 20.0$, $M_{12} = 23.1$, $M_{24} = 28.2$ and M_{36} = 36.8, where the subscripts refer to thc value of X. The output that follows is an edited version of a linear regression analysis produced by *SYSTAT*.

```
Dep Var: Y  N: 40   Multiple R: 0.8586   Squared multiple R: 0.7373
Effect          Coefficient      Lower  < 95%>  Upper

CONSTANT          18.7000       16.6645        20.7355
X                  0.4625        0.3718         0.5532
```

```
                         Analysis of Variance
Source        Sum-of-Squares  df    Mean-Square      F-ratio        P

Regression        1540.1250    1     1540.1250     106.6316     0.0000
Residual           548.8500   38       14.4434
```

The first part of the output tells us that point estimates of the values of the parameters of the linear regression model are $\hat{\beta}_0 = 18.70$ and $\hat{\beta}_1 = 0.4625$, so the sample regression equation is

$$Y = 18.70 + 0.4625X,$$

the equation of the straight line of best fit shown in Figure D1. (A line of best fit minimizes the sum of squares of the vertical distances between the sample means and the line.) This straight regression line represents the *linear component of trend* in the relationship between X values and Y means. In this particular case the best-fitting straight line has a positive slope, suggesting that there is a positive linear relationship between X and Y. In addition, the sample means are all reasonably close to the line, suggesting that *nonlinear* components of trend (systematic departures from linearity in the relationship between X and Y) are likely to be relatively small and perhaps trivial.

The constant β_0 is of less interest than the value of the regression coefficient β_1, which can be interpreted as the size of the effect on Y of a unit increase (an increase of 1.0) in X. $\hat{\beta}_1 = 0.4625$ is a point estimate of β_1. The output also includes the 95% CI $\beta_1 \in (0.372, 0.553)$. We infer from this CI that an increase of one hour in the amount of sleep deprivation results in an increase of at least

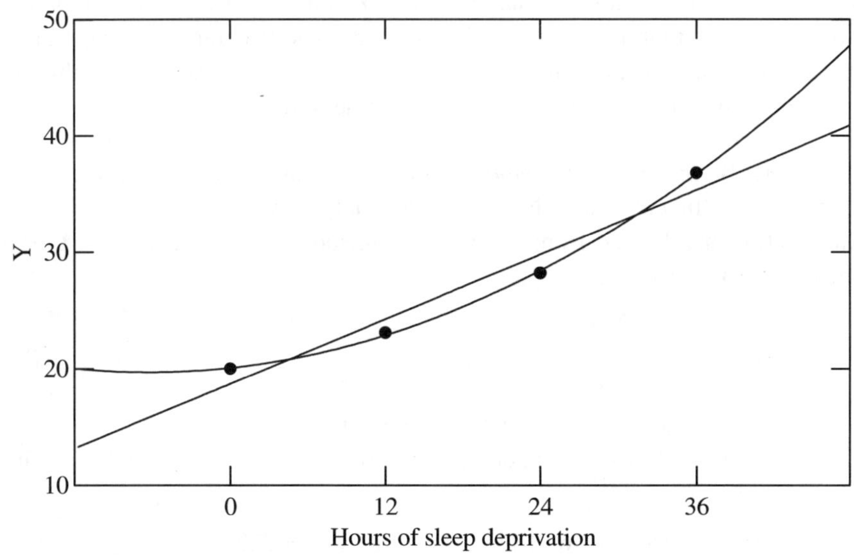

Figure D1 *Means and linear and quadratic regression lines*

0.372 (but not more than 0.553) errors on the driving simulator task This is a reasonable inference provided that the model fits the data well and the equation is not applied outside the range of *X* values (sleep deprivation levels) included in the experiment.

If dependent variable units (errors on the task) are uninformative, we can transform this raw interval into an approximate standardized interval by dividing the interval limits by the square root of the *mean square residual* figure given in the last section of the output. The approximate standardized interval is $\beta_1/\sigma_\varepsilon \in (0.104, 0.155)$, where σ_ε is the standard deviation of the error variable defined by the linear regression model. Thus, according to this analysis, each additional hour of sleep deprivation (in the range of 0 to 36 hours) produces an increase in error score of a little over one-tenth of a standard deviation.

It would be a mistake to apply Cohen's (1988) effect size guidelines to this result, thereby concluding that sleep deprivation has a small effect on performance on this particular task. The CI on $\beta_1/\sigma_\varepsilon$ implies that $36\beta_1/\sigma_\varepsilon \in (3.744, 5.580)$, so that, according to the analysis, 36 hours of sleep deprivation has a very large effect. The quantitative nature of the independent variable reminds us that effect size is a function of what Abelson (1995) calls *cause size*.

A section of the output not shown here includes a point estimate (0.859) of the *standardized regression coefficient* (under the heading *Std Coef* in *SYSTAT* output, called *beta* in *SPSS* output). It is important to recognize that this is not an estimate of the parameter $\beta_1/\sigma_\varepsilon$. The standardized regression coefficients reported in regression analyses refer to standardized versions of both independent and dependent variables (*X* and *Y*), and the standardization of *Y* is based on total variation, not variation within groups. It is rarely appropriate to standardize independent variables in fixed-effects experiments, where values of *X* are selected by the experimenter rather than measured.

Estimating β_1 *in an ANOVA-model analysis* In order to estimate the linear regression coefficient in an ANOVA-model analysis, we must define a contrast with coefficients derived from the *X* values included in the experiment. Linear trend contrast coefficients are

$$c_j = X_j - \frac{\sum_j X_j}{J} \qquad \text{(D2)}$$

where X_j is the value of *X* associated with treatment *j*.

The X_j values in this experiment are $X_1 = 0$, $X_2 = 12$, $X_3 = 24$ and $X_4 = 36$, so $\sum X_j / J = 72/4 = 18$ and

$$\mathbf{c}'_{lin} = [-18 \ \ -6 \ \ 6 \ \ 18] \qquad (\sum c_{lin}^2 = 720).$$

The population value of the linear regression coefficient is

$$\beta_1 = \frac{\psi_{lin}}{\sum c_{lin}^2} \tag{D3}$$

whether or not intervals between adjacent X values are equal, and whether or not samples sizes are equal.

If *PSY* is used to construct a CI on ψ_{lin}, the *No Rescaling* option should be chosen. The following (edited) output is from an analysis that treats the linear trend contrast is the only contrast in a planned analysis.

```
                    Individual 95% Confidence Intervals
                    -----------------------------------

                    Raw CIs (scaled in Dependent Variable units)
                    -----------------------------------------------------
                    Contrast    Value         SE          ..CI limits..
                                                         Lower        Upper
                    -----------------------------------------------------
Linear              B1        333.000      30.728      270.681      395.319
                    -----------------------------------------------------

                 Approximate Standardized CIs (scaled in Sample SD units)
                    -----------------------------------------------------
                    Contrast    Value         SE          ..CI limits..
                                                         Lower        Upper
                    -----------------------------------------------------
Linear              B1         91.956       8.485       74.747      109.165
                    -----------------------------------------------------
```

To transform these statistics into statistics referring to the linear regression coefficient, we need to divide by $\sum c_{lin}^2 = 720$. Thus the point estimate of the raw regression coefficient is

$$\hat{\beta}_1 = \frac{333.0}{720} = 0.4625,$$

the sample regression coefficient obtained from the linear regression analysis. This is the only statistic that reproduces exactly some part of the output of a regression analysis, due to differences in the definition of standard errors by the unsaturated regression model and a saturated ANOVA or means model.

An exact standardized CI on $\beta_1/\sigma_\varepsilon$ (where σ_ε is defined by a saturated model) can be obtained from *STATISTICA Power Analysis* as follows:

- provide the program with the t ($= \sqrt{F}$) statistic for ψ_{lin} obtained from *PSY* (10.8371 in this case), together with $\sum c_{lin}^2 = 720$, and
- divide the limits of the resulting CI by $\sum c_{lin}^2 = 720$.

The resulting exact 95% CI is $\beta_1/\sigma_\varepsilon \in (0.090, 0.165)$.

Nonlinear components of trend

A saturated *polynomial regression model* has $(J - 1)$ independent variables, each of which is a power of X: $X^1 (= X)$, X^2, $X^3, \ldots, X^{J-1}$. Most polynomial regression analyses are based on one or more *unsaturated* models which include only X or perhaps X and X^2 as independent variables. If $J = 4$, potential polynomial regression models are

Linear regression model $\qquad Y = \beta_0 + \beta_1 X + \varepsilon \qquad$ (D4)

Quadratic regression model $\qquad Y = \beta_0 + \beta_1 X + \beta_2 X^2 + \varepsilon \qquad$ (D5)

Cubic regression model $\qquad Y = \beta_0 + \beta_1 X + \beta_2 X^2 + \beta_3 X^3 + \varepsilon. \qquad$ (D6)

Each model provides a better fit than the previous model in the hierarchy to the set of population means (μ_1 to μ_4), unless some model simpler than the saturated model (D6) fits the means perfectly.

The parameters of a polynomial regression model are the constant β_0 and the regression coefficients ($\beta_1, \beta_2, \ldots$) referring to powers of X. In general, none of the parameters of a relatively complex model is identical to any of the parameters in simpler models. In particular, the linear regression coefficient β_1 defined by (D4) is not identical to the β_1 parameters defined by (D5) or (D6).

The quadratic component of trend The β_2 coefficient in the quadratic regression model (D5) quantifies the degree of change in the size of the effect of X on Y as X increases. Specifically, $2\beta_2$ is the increase in the size of the effect of a unit increase in X. Suppose, for example, that the population means are described perfectly by a quadratic model with parameters $\beta_0 = 20$, $\beta_1 = 0.1$ and $\beta_2 = 0.01$; that is, by the quadratic model equation

$$Y = 20 + 0.1X + 0.01X^2. \tag{D7}$$

According to this model, if X is increased by 1 hour of sleep deprivation when the initial value is $X = 0$, then Y increases from 20 to 20.11, an increase of 0.11. If X is increased by one more hour (from 1 to 2), then Y increases from 20.11 to 20.24, an increase of 0.13. The difference in the effects of these adjacent increases of one hour in X is $0.13 - 0.11 = 0.02 = 2\beta_2$. The same rate of increase in the effect of X applies throughout the relevant range of X values (0 to 36). If X is increased from 34 to 35 hours, the increase in Y is 0.79. If X is increased by one more hour (to 36 hours), the increase in Y is 0.81. Again, the difference between these adjacent effects $(0.81 - 0.79)$ is 0.02.

The average of the smallest (0.11) and largest (0.81) of the effects of a unit increase in X in the relevant range of X values is equal to the regression coefficient β_1 in the *linear* regression model equation

$$Y = 18.56 + 0.46X.$$

The value $\beta_1 = 0.1$ in (D7) cannot be readily interpreted.

In general, if a quadratic regression model fits the population means perfectly, the regression coefficient from a linear regression model (given equal intervals between adjacent X levels) is the *average* effect of a unit increase in X over the range of X values in the experiment.

Estimating β_2 In order to estimate β_2 it is necessary to carry out a multiple regression analysis with X and X^2 as independent variables. When applied to the sleep deprivation data, *SYSTAT* produces the following point and interval estimates of the parameters of the quadratic regression model (D5).

Effect	Coefficient	Lower	< 95%>	Upper
CONSTANT	20.075000	17.841554		22.308446
X	0.118750	-0.180143		0.417643
X2	0.009549	0.001592		0.017505

The curved line in Figure D1 is a graph of the sample regression equation

$$Y = 20.075 + 0.11875X + 0.009549X^2.$$

The quadratic regression line fits the sample means better than the linear regression line, as it must unless the two equations are identical (which can only happen if $\hat{\beta}_2 = 0$ in the quadratic-model analysis). It is possible, of course, that the quadratic model might provide only a trivial improvement in fit, at the cost of an increase in complexity.

The fact that the vertical distance between the linear and quadratic regression lines in Figure D1 increases when they are extrapolated beyond the range of X values in the data is typical of polynomial regression lines of different degree. It may be possible to justify interpolation within the range of X values in the data, but polynomial regression analysis provides no justification for extrapolation.

Orthogonal polynomial contrasts in ANOVA-model analyses When sample sizes are equal and intervals between adjacent levels of X are also equal, it is possible to avoid the complexities of polynomial regression models by specifying a set of *orthogonal polynomial contrasts* in an ANOVA-model analysis. Each orthogonal polynomial contrast refers to the highest-order trend component defined by a particular polynomial regression model. Tables of orthogonal polynomial contrast coefficients have been widely published (by Kirk, 1995, and Winer, Brown and Michels, 1991, among others) and trend analyses based on significance tests are supported by most statistical packages.

The main problem with standard orthogonal polynomial contrast analyses (apart from the restrictions noted in the previous paragraph) is the absence of any provision for rescaling the contrast coefficients so that the magnitude of contrast values can be interpreted.

Appendix E Solutions

Chapter 1

1. (a) (i) (Confidence interval inference): The amount of practice given in the experiment has a nontrivial positive effect, increasing the average score on the aptitude test by at least 6.5 (but not more than 8.7) items correct.

(ii) (Confident direction inference): Practice increases the average score.

(iii) (Confident inequality inference): Practice has an effect on the average score.

(b) (i) Practice increases the average score on the aptitude test. We cannot be confident, however, about the magnitude of the effect, which may be trivially small (as low as 0.9) or substantial and important (as high as 16.8).

(ii) Practice increases the average score.

(iii) Practice has an effect on the average score.

(c) (i) The amount of practice given in the experiment has a trivially small effect of unknown sign, and is practically equivalent to the effect of answering questions in an interest inventory.

(ii) No confident directional inference is possible.

(iii) No confident inequality inference is possible.

(d) (i) The experiment has produced such an imprecise estimate of the size of the practice effect that it provides almost no information. The effect size might be nontrivial and negative (it might decrease the average test score by as much as 7.4 items correct), substantial and positive (it might increase the average score by as much as 8.5), or anything in between.

(ii) Confident directional inference is not possible.

(iii) Confident inequality inference is not possible.

2. (a) $\mu_T - \mu_C \in (0.79\sigma, 1.06\sigma)$ [or $(\mu_T - \mu_C)/\sigma \in (0.79, 1.06)$]. By conventional standards this is a large effect. Practice has a substantial (close to 1 standard deviation in magnitude) positive effect on average test score.

(b) $\mu_T - \mu_C \in (0.11\sigma, 2.05\sigma)$. Although we can be confident that practice increases the average score on the aptitude test, we know almost nothing about the magnitude of the effect. The effect size might differ only trivially from zero, it might be very large (up to about 2 standard deviations), or it might be anywhere between these extremes.

(c) $\mu_T - \mu_C \in (-0.07\sigma, 0.20\sigma)$. By conventional standards, the amount of practice given in the experiment produces a very small effect of unknown sign (perhaps no effect). If the smallest nontrivial effect is taken to be 0.37 standard deviation units (from Question 1), then it can be concluded that the effect of the practice given on test performance is practically equivalent to the effect of answering questions in the interest inventory.

(d) $\mu_T - \mu_C \in (-0.90\sigma,\ 1.04\sigma)$. The experiment provides almost no useful information. The effect size might be substantial and negative, substantial and positive, or anything in between. All we can infer is that the effect is not massively large (greater than about 1 standard deviation).

3. The half-widths of the standardized confidence intervals are (a) 0.135, (b) 0.97, (c) 0.135 and (d) 0.97. Intervals (a) and (c) are very narrow, so they provide very precise estimates of $\mu_T - \mu_C$. Intervals (b) and (d) are very wide, so they provide relatively imprecise estimates. The sample sizes on which intervals (a) and (c) are based must have been equal to each other (assuming equal *n*s in each case) and very large, relative to sample sizes usually employed in experimental psychology. The sample sizes on which intervals (b) and (d) are based must have been equal to each other and relatively small.

4. Replication 1: There are no inferential errors at any level (the confidence interval covers $\mu_1 - \mu_2$, there is no directional inference and no inequality inference). Replication 18: There is a noncoverage error (the entire interval is above $\mu_1 - \mu_2$, but there are no errors at lower levels (the directional inference $\mu_1 > \mu_2$ and the inequality inference $\mu_1 \neq \mu_2$ are both correct).

5. No. The experimenter would not know the value of $\mu_1 - \mu_2$, and could not know whether any inference was in error. (The experimenter would know, of course, that if no inference was made, then no erroneous inference could have been made.)

Chapter 2

1. (a) The required contrast coefficient vectors are

[1.00 –0.25 –0.25 –0.25 –0.25]
[0 0.5 –0.5 0.5 –0.5]
[0 0.5 0.5 –0.5 –0.5]
[0 1 –1 –1 1].

Note that the last of these coefficient vectors does not define a mean difference contrast. The contrast can be written as $\psi_4 = (\mu_2 - \mu_4) - (\mu_3 - \mu_5)$, the difference between the size of the contingency effect when the fee is \$100 and the size of the contingency effect when the fee is \$200.

(b) Individual confidence intervals on the planned contrasts are

(7.10, 8.81)
(1.89, 3.41)
(7.54, 9.07)
(–2.03, 1.03).

(If you are not sure where these come from, consult the *PSY* output file *Ch2 Q1b.out*.) We can conclude that the average effect of charging a fee (averaging across the four fee conditions in the experiment) is a clinically important increase in the effectiveness of the treatment. The fee effect is substantially greater when the amount of the fee does not depend on treatment outcome. A fee of \$200 produces a slightly better outcome than a fee of \$100, but this is not a clinically significant difference. The size of the contingency effect depends only trivially (if at all) on the size of the fee.

(c) The average effect of charging a fee is an increase in the effectiveness of the treatment. The fee effect is greater when the amount of the fee does not depend on treatment outcome. A fee of $200 produces a better outcome than a fee of $100. No inference can be drawn about whether the size of the contingency effect varies across fee levels.

(d) Simultaneous confidence intervals on the planned contrasts are

(i) (6.85, 9.05)
(ii) (1.67, 3.63)
(iii) (7.32, 9.28)
(iv) (–2.46, 1.46).

(These intervals are obtained by selecting *Bonferroni t* from the *Analysis Options* menu.) The simultaneous intervals are slightly wider than the corresponding individual intervals, but this slight decrease in precision has no effect on the conclusions drawn in this particular case.

(e) The sample mean vector is $\mathbf{m}' = [\ 18.65\ 14.05\ 13.65\ 14.00\ \ 5.10\ ]$.
Bonferroni-*t* SCIs on the planned contrasts are

(i) (5.75, 8.15)
(ii) (3.57, 5.73)
(iii) (3.22, 5.38)
(iv) (–10.65, –6.35).

We can conclude that the average effect of charging a fee (averaging across the four fee conditions in the experiment) is to produce a clinically important increase in the effectiveness of the treatment. Averaging across the two fee levels included in the experiment, the magnitude of the fee level effect is greater when the requirement of a fee does not depend on treatment outcome. Averaging across the two contingency levels, a fee of $200 produces a better outcome than a fee of $100. It is not clear whether either of these average effects is clinically significant, although both may be. There is a substantial difference between the two fee levels in the size of the contingency effect, such that the desirable effect of making the fee nonrefundable is greater for the $200 fee than for the $100 fee.

This analysis provides a rather convoluted account of what appears to have happened in the experiment, mainly because inferences on both of the average effects (the average fee effect and the average contingency effect) are somewhat uninformative. In both cases it appears (after the event) that a clearer account would follow from an analysis of individual rather than average effects, because of the substantial differences in individual effects shown by the final contrast. The set of contrasts planned for the analysis makes no provision for direct inferences on individual effects.

The same set of planned contrasts provided a more satisfying analysis of the data used in (d) because it turned out that inferences on average effects implied informative inferences on individual effects, given the implications of the confidence interval on the last contrast. That interval implied that the two fee size effects (one for each contingency) and the two contingency effects (one for each fee) were practically equivalent.

(f) This question can be answered in a number of different ways, because of the flexibility of a post hoc analysis. The analysis reported here includes all of the contrasts used in the planned analyses [see 1(a) above], as well as the following contrasts dealing with individual fee level or contingency effects:

(v) [0 1 –1 0 0] (fee level effect, given possibility of refund)
(vi) [0 0 0 1 –1] (fee level effect, given no refund possible)
(vii) [0 1 0 –1 0] (contingency effect, given \$100 fee)
(viii) [0 0 1 0 –1] (contingency effect, given \$200 fee).

Simultaneous confidence intervals on these contrasts in a post hoc analysis (from the file *Ch2 Q1f.out*) are

(i) (5.60, 8.30)
(ii) (3.44, 5.86)
(iii) (3.09, 5.51)
(iv) (–10.92, –6.08)
(v) (–1.31, 2.11)
(vi) (7.19, 10.61)
(vii) (–1.66, 1.76)
(viii) (6.84, 10.26).

While the first four of these intervals are slightly wider than the corresponding intervals in the planned analysis [in 1(e)], they do not lead to different conclusions. We can conclude from the additional post hoc contrasts that the two fee levels have practically equivalent effects when there is the possibility of a refund, but a \$200 fee produces a substantially better outcome than a \$100 fee when no refund is possible. Similarly, the two contingency conditions are practically equivalent when the fee is \$100, but the possibility of an outcome-dependent refund substantially reduces the effect of the \$200 fee.

This analysis seems to be more informative than the planned analysis in (e) because, given the way the data turned out, the additional contrasts [(v) to (viii)] provided clear answers to the questions of interest to the experimenter. Had the data turned out differently [as in (b)], the planned analysis might have been satisfactory.

(g) The justification for the claim that Bonferroni-t confidence intervals (or tests) control the FWER depends on the assumption that direct inferences will be made only on those k contrasts planned for the analysis independently of the data. Sometimes, however, experimenters feel free to accept the benefits of a restricted analysis (namely increased precision) when they like the outcome of the analysis, while also feeling free to take advantage of the flexibility of an unrestricted analysis when they do not like the outcome of the restricted analysis. This post hoc selection of analysis strategy (as distinct from a post hoc choice of contrasts in an unrestricted analysis) capitalizes on chance and inflates the FWER.

It is not clear, however, that the inflation of the FWER associated with the practice of following a 'failed' planned analysis with a post hoc analysis is a particularly serious problem in practice. The FWER associated with this approach cannot exceed (and would always be somewhat less than) 2α.

Chapter 3

1. (a) From the first column in Table F2 or F4, $n = 88$ when $w = 0.25$ and $\alpha = .10$. This applies to the comparison, so $w < .25$ for the confidence interval on the {2, 1} contrast. The overall sample size required is $3 \times 88 = 264$.

(b) The Tukey procedure should be used for this analysis. From Table F2, $n = 136$, so $N = 3 \times 136 = 408$.

(c) The Bonferroni-t procedure is appropriate here. From Table F3, n = 124, so $N = 3 \times 124 = 372$.

(d) The Scheffé procedure is required for this analysis. From Table F4, n = 149, so $N = 3 \times 149 = 447$.

2. (a) The Scheffé procedure can produce 90% simultaneous confidence intervals on any contrasts of interest. Table F7 shows that $\lambda = 3.175$ when $\nu_1 = J - 1 = 3$, $\alpha = .10$ and $(1 - \beta) = .75$. Given an effect size of $\gamma = 0.8$, we can use (3.9) to solve for the n required to produce a conditional power figure for comparisons of $(1 - \beta) = .75$:

$$n = 1 + 2\left(\frac{3.175}{0.8}\right)^2 = 32.5.$$

Thus an n of 33 subjects per group ($N = 4n = 132$) will provide the required conditional probability of directional inference from confidence intervals on comparisons. This sample size will ensure that the power of a Scheffé test on any $\{m, r\}$ contrast must be at least .75 when $\alpha = .10$ and $\gamma = 0.8$.

(b) An analysis restricted to comparisons should be carried out with the Tukey procedure. Table F5 shows that $\lambda = 2.966$ when $J = 4$, $\alpha = .10$ and $(1 - \beta) = .75$. The required n is therefore

$$n = 1 + 2\left(\frac{2.966}{0.8}\right)^2 = 28.5.$$

Rounding up, $n = 29$ and $N = 4n = 116$.

(c) The Scheffé analysis allows for direct inferences on {2, 2}, {3, 1} and {2, 1} contrasts as well as comparisons. Not only is this analysis more flexible, but it also provides greater conditional power for tests on these contrasts than for tests on comparisons. Applying (3.9) to a {2, 2} contrast with $\sum c^2 = 1$ shows that the required power can be achieved with a much smaller n (n = 17, N = 68), so if n = 33 and $\gamma = .8$ the probability of a directional inference on this type of contrast must be considerably higher than .75. The advantage of the Tukey analysis is that the sample size required to achieve the required power for tests on comparisons is only 88% as large as that required by the Scheffé analysis.

Chapter 4

1. (a) No. The difference between $A(b_1)$ and $A(b_2)$ is the AB interaction contrast (appropriately scaled), and this contrast is not included in the analysis. The outcome of the tests carried out by the experimenter justify only the inference $A(b_1) > 0$, implying nothing about the difference between the two simple effect contrasts.

(b) Individual CIs from the *PSY* analysis are shown below.

```
Raw CIs (scaled in Dependent Variable units)
-------------------------------------------------------
Contrast         Value        SE          ..CI limits..
                                         Lower       Upper
-------------------------------------------------------
A(b1)            2.000       0.656       0.704       3.296
A(b2)            1.000       0.656      -0.296       2.296
B                3.500       0.464       2.583       4.417
-------------------------------------------------------
```

Approximate Standardized CIs (scaled in Sample SD units)

Contrast	Value	SE	..CI limits.. Lower	Upper
A(b1)	0.681	0.224	0.240	1.123
A(b2)	0.341	0.224	-0.101	0.782
B	1.192	0.158	0.880	1.505

(i) The *A* effect at b_1 is positive, and may be small, very large, or somewhere in between. If the *A* effect at b_2 is positive it may be substantial, trivially small, or anywhere in between. If this effect is negative it is trivially small.

(ii) If $A(b_1) > A(b_2)$, the difference between them may be as large as 1.12 – (–0.10) = 1.22 standard deviation units. If $A(b_1) < A(b_2)$, the difference between them may be as large as 0.78 – 0.24 = 0.54 standard deviation units.

(c) The difference between the two *A* simple effects is the interaction contrast *AB*, so the analysis should provide for inferences on all contrasts concerned with differences between levels of [*A*, $A(b_1)$, $A(b_2)$ and *AB*] and on the *B* main effect. This can be achieved by defining two families of contrasts, *A*(*B*) and *B*. The PFER for the *A*(*B*) family can be set at $2\alpha = .10$, or the FWER can be set at $1-(1-\alpha)^2 = .0975$. The *A*(*B*) family has $df_A + df_{AB} = 2\ df$ and includes four planned contrasts. The CC for this family should be the smaller of $t_{.10/(2\times4);156} = 2.2632$ and $\sqrt{2F_{.0975;\,2,156}} = 2.1739$. The analysis requires three *PSY* runs, one to produce CIs for the mean difference contrasts in the *A*(*B*) analysis, one to provide scaling appropriate for the interaction contrast in that analysis, and a third to construct an individual 95% CI on the *B* main effect contrast.

The *PSY* output file is *Ch4 Q1c.out*. Edited CI tables follow.

Raw CIs (scaled in Dependent Variable units)

Contrast	Value	SE	..CI limits.. Lower	Upper
A	1.500	0.464	0.491	2.509
AB	1.000	0.928	-1.018	3.018
A(b1)	2.000	0.656	0.573	3.427
A(b2)	1.000	0.656	-0.427	2.427
B	3.500	0.464	2.583	4.417

Approximate Standardized CIs (scaled in Sample SD units)

Contrast	Value	SE	..CI limits.. Lower	Upper
A	0.511	0.158	0.167	0.855
AB	0.341	0.316	-0.347	1.028
A(b1)	0.681	0.224	0.195	1.167
A(b2)	0.341	0.224	-0.145	0.827
B	1.192	0.158	0.880	1.505

The CI on *AB* suggests that if $A(b_1) > A(b_2)$, the difference between them may be as large as 1.03 standard deviation units. If $A(b_1) < A(b_2)$, the difference between them is no larger than 0.35 standard deviation units.

Chapter 5

1. (a) (i) A is the average drug effect (medication – placebo) on the dependent variable, averaged across the three psychological treatments.

(ii) If we define the (CBT – C) difference as the effect of CBT and the (PT – C) difference as the effect of PT, then B_1 is the average of these two effects, averaging also across the drug treatment and placebo conditions.

(iii) AB_2 is the difference between the drug and placebo conditions in the size of the difference between the CBT and PT means. It is also the difference between the CBT and PT conditions in the size of the drug effect.

(b) The CC for the single A main effect contrast is $t_{.05/2;114} = 1.981$. The B main effect has 2 *dfs* and four planned contrasts. The CC should be the smaller of $t_{.05/(2\times4);114} = 2.538$ and $\sqrt{2F_{.05;2,114}} = 2.480$. The same is true of the AB interaction effect. Therefore the Scheffé CC of 2.480 should be used for B and AB contrasts.

(c) (i) The $B(A)$ family includes all of the planned B main and simple effect contrasts, as well as the planned AB interaction contrasts, but does not include the A main effect contrast. An exhaustive analysis would include two families: A, with an FWER (equivalent in this case to a PCER) of $\alpha = .05$, and $B(A)$, with a PFER of $2\alpha = .10$ or an FWER of $1-(1-\alpha)^2 = .0975$. The $B(A)$ family has $df_A + df_{AB} = 4$ *df* and includes 16 planned contrasts: four B main effect, four AB interaction effect and eight B simple effect contrasts. The CC for this family should be the smallest of $t_{.10/(2\times16);114} = 2.786$, $\sqrt{4F_{.0975;4,114}} = 2.837$ and $\sqrt{SMR_{.0975;2,2,114}} = 2.692$. The SMR CC should be used for the $B(A)$ family, whether or not the A main effect contrast (evaluated with a CC of 1.981) is of any interest to the experimenter in an analysis excluding A simple effect contrasts.

(ii) The contrasts section of the *PSY* input file should be something like this:

```
[BetweenContrasts]
1  1  1 -1 -1 -1 A
1  1 -2  1  1 -2 B1
1 -1  0  1 -1  0 B2
1  0 -1  1  0 -1 B3
0  1 -1  0  1 -1 B4
1  1 -2  0  0  0 B1(a1)
0  0  0  1  1 -2 B1(a2)
1 -1  0  0  0  0 B2(a1)
0  0  0  1 -1  0 B2(a2)
1  0 -1  0  0  0 B3(a1)
0  0  0  1  0 -1 B3(a2)
0  1 -1  0  0  0 B4(a1)
0  0  0  0  1 -1 B4(a2)
1  1 -2 -1 -1  2 AB1
1 -1  0 -1  1  0 AB2
1  0 -1 -1  0  1 AB3
0  1 -1  0 -1  1 AB4
```

2. (a) All raw confidence intervals cover the relevant parameters, so there are no noncoverage errors.

(b) All approximate standardized confidence intervals cover the relevant parameters, so there are no noncoverage errors.

3. (a) Each factor has two levels, so each main effect has 1 *df*. There is a first-order (double) interaction for each pair of factors, a second-order (triple) interaction for each

combination of three factors, and one third-order (quadruple) interaction involving all four factors. Degrees of freedom for interactions are obtained by multiplication. Because the model is saturated, the sum of the *df*s for effects should be the number of degrees of freedom between cells: $2 \times 2 \times 2 \times 2 - 1 = 15$. Effects and *df*s are

Effect	*df*
A	1
B	1
C	1
D	1
AB	1
AC	1
AD	1
BC	1
BD	1
CD	1
ABC	1
ABD	1
ACD	1
BCD	1
ABCD	1
Between cells	15

(b) There are $16(n - 1) = 144$ degrees of freedom for error. Because there is only one contrast per 'family', the CC for all families (and therefore all contrasts in the analysis) is $t_{.05/2;144} = 1.977$.

(c) (i) Assuming that the 16 groups (cells) in the data section of the *PSY* input file are ordered so that the levels of *A* change most slowly and the levels of *D* change most quickly, the contrasts section of the file should look like this:

```
[BetweenContrasts]
1  1  1  1  1  1  1  1 -1 -1 -1 -1 -1 -1 -1 -1 A
1  1  1  1 -1 -1 -1 -1  1  1  1  1 -1 -1 -1 -1 B
1  1 -1 -1  1  1 -1 -1  1  1 -1 -1  1  1 -1 -1 C
1 -1  1 -1  1 -1  1 -1  1 -1  1 -1  1 -1  1 -1 D
1  1  1  1 -1 -1 -1 -1 -1 -1 -1 -1  1  1  1  1 AB
1  1 -1 -1  1  1 -1 -1 -1 -1  1  1 -1 -1  1  1 AC
1 -1  1 -1  1 -1  1 -1 -1  1 -1  1 -1  1 -1  1 AD
1  1 -1 -1 -1 -1  1  1  1  1 -1 -1 -1 -1  1  1 BC
1 -1  1 -1 -1  1 -1  1  1 -1  1 -1 -1  1 -1  1 BD
1 -1 -1  1  1 -1 -1  1  1 -1 -1  1  1 -1 -1  1 CD
1  1 -1 -1 -1 -1  1  1 -1 -1  1  1  1  1 -1 -1 ABC
1 -1  1 -1 -1  1 -1  1 -1  1 -1  1  1 -1  1 -1 ABD
1 -1 -1  1  1  1 -1 -1 -1  1  1 -1 -1  1  1 -1 ACD
1 -1 -1  1 -1  1  1 -1  1 -1 -1  1 -1  1  1 -1 BCD
1 -1 -1  1 -1  1  1 -1 -1  1  1 -1  1 -1 -1  1 ABCD
```

The coefficients for all interaction contrasts are obtained by multiplication.

(ii) Four. It would be necessary to make one run with all default options in place to produce individual confidence intervals with mean difference scaling appropriate for the four main effect contrasts, a second run to produce scaling appropriate for first-order interaction contrasts (*AB*, *AC*, *BC*, *BD* and *CD*), a third run to produce scaling appropriate for second-order interaction contrasts (*ABC*, *ABD*, *ACD* and *BCD*) and a final run to produce scaling appropriate for the third-order interaction contrast *ABCD*.

Chapter 6

1. (a) The raw CIs (from the *PSY* output file *Ch6 Q1a.out*) are

Contrast	Value	SE	..CI limits.. Lower	Upper
A (SD)	11.980	0.387	11.182	12.778
B (NVH)	3.300	0.439	2.394	4.206
AB	-10.680	0.979	-12.701	-8.659

Sleep deprivation (SD) produces a very large increase in the error score (at least 11.2 dependent variable units) averaged across NVH levels. Averaged across SD levels, high NVH levels also produce a nontrivial increase in the error score (at least 2.4), but the NVH main effect (which is no larger than 4.2) is much smaller than the SD main effect. The level of NVH has a very large influence on the size of the SD effect. It is clear from the *AB* confidence interval that if SD increases the error score at both NVH levels, then that effect is much smaller (by at least 8.7) at the high NVH level than at the low NVH level.

(b) This analysis is carried out by selecting the post hoc option and setting the confidence level at $100(1-.05)^3 = 85.74\%$. The raw CIs (from the *PSY* output file *Ch6 Q1b.out*) are

Contrast	Value	SE	..CI limits.. Lower	Upper
A (SD)	11.980	0.387	10.989	12.971
B (NVH)	3.300	0.439	2.176	4.424
AB	-10.680	0.979	-13.189	-8.171
A(b1)	6.640	0.629	5.027	8.253
A(b2)	17.320	0.618	15.736	18.904
B(a1)	-2.040	0.662	-3.736	-0.344
B(a2)	8.640	0.653	6.967	10.313

The *A*, *B* and *AB* confidence intervals are marginally wider than those in the previous analysis, but this slight decrease in precision has virtually no effect on the interpretations of these intervals. The simple effect contrasts make it clear that SD produces a substantial increase in errors at both NVH levels (at least 5.0 at the high NVH level and at least 15.7 at the low NVH level). A high NVH level has a small (perhaps trivially small) beneficial effect on sleep-deprived subjects, but a substantial detrimental effect (producing an increase in error scores of at least 7.0) on those who are not sleep deprived. These inferences cannot be obtained from the standard analysis.

2. (b) This analysis is carried out by selecting the post hoc option and setting the confidence level at $100(1-.05)^3 = 85.74\%$. The approximate standardized confidence intervals are

Contrast	Value	SE	..CI limits.. Lower	Upper
A (SD)	11.980	0.960	10.074	13.886
B (NVH)	3.300	0.960	1.394	5.206
AB	-10.680	1.921	-14.493	-6.867
A(b1)	6.640	1.358	3.944	9.336
A(b2)	17.320	1.358	14.624	20.016
B(a1)	-2.040	1.358	-4.376	0.656
B(a2)	8.640	1.358	5.944	11.336

Unlike the within-subjects analysis, this analysis does not justify a confident inference about the direction of the effect of NVH on sleep-deprived subjects. Further, this analysis leaves open the possibility that the NVH main effect may be trivially small. For the most part, though, inferences from this analysis are very similar to those from the within-subjects analysis.

Chapter 7

1. (a) One of the 11 raw confidence intervals does not cover the population value of the relevant contrast. The within-subjects main effect contrast W_2, the difference between the Post mean and the average of the follow-up means, has a population value of –3.5, so the confidence interval inference

$$W_2 \in (-5.668, -3.621)$$

is in error. This particular error would have no substantive implications for the interpretation of the data, partly because the upper limit of the interval is very close to the population value, and partly because within-subjects main effect contrasts are not particularly important in this study.

(b) Each of the 12 approximate standardized confidence intervals covers the population value of the relevant contrast, so there are no noncoverage errors.

(c) All of the contrasts standardized on the basis of pre-treatment variation have slightly larger population values than when standardization is based on the square root of the average of population variances, because the pre-treatment variance is smaller than all subsequent variances.

2. The CC for all between-subjects effects is calculated from

$$\sqrt{\nu_B F_{\alpha;\nu_B,\nu_E}} \quad \text{(7.8, repeated)}$$

For the A effect (with $\nu_B = 1$), $\text{CC} = \sqrt{F_{.05;1,54}} = 2.005$.

For the B and AB effects (with $\nu_B = 2$), $\text{CC} = \sqrt{2F_{.05;2,54}} = 2.517$.

The CC for all within-subjects effects is calculated from

$$\sqrt{\frac{\nu_W \nu_E}{\nu_E - \nu_W + 1} F_{\alpha;\nu_W,\nu_E-\nu_W+1}} \quad \text{(7.9, repeated)}$$

For the (*C*) and *A*(*C*) effects (with $\nu_W = 3$), $\text{CC} = \sqrt{\frac{3\times 54}{52} F_{.05;3,52}} = 2.944$.

For the (*D*) and *A*(*D*) effects (with $\nu_W = 4$), $\text{CC} = \sqrt{\frac{4\times 54}{51} F_{.05;4,51}} = 3.289$.

For the (*CD*) and *A*(*CD*) effects (with $\nu_W = 12$), $\text{CC} = \sqrt{\frac{12\times 54}{43} F_{.05;12,43}} = 5.470$.

Note that for the purpose of calculating the CC, the *A*(*C*), *A*(*D*) and *A*(*CD*) effects are treated as within-subjects effects because $\nu_B = 1$.

The CC for all between × within effects is calculated from

$$\sqrt{\frac{\nu_E \theta_{\alpha;s,m,n}}{1-\theta_{\alpha;s,m,n}}} \qquad (7.10, \text{repeated})$$

where $s = \min(\nu_B, \nu_W)$, $m = (|\nu_B - \nu_W| - 1)/2$ and $n = (\nu_E - \nu_W - 1)/2$.

For the *B*(*C*) and *AB*(*C*) effects (with $\nu_B = 2$ and $\nu_W = 3$),

$$\text{CC} = \sqrt{\frac{54\theta_{.05;2,0,25}}{1-\theta_{.05;2,0,25}}} = 3.484.$$

For the *B*(*D*) and *AB*(*D*) effects (with $\nu_B = 2$ and $\nu_W = 4$),

$$\text{CC} = \sqrt{\frac{54\theta_{.05;2,0.5,24.5}}{1-\theta_{.05;2,0.5,24.5}}} = 3.842.$$

For the *B*(*CD*) and *AB*(*CD*) effects (with $\nu_B = 2$ and $\nu_W = 12$),

$$\text{CC} = \sqrt{\frac{54\theta_{.05;2,4.5,20.5}}{1-\theta_{.05;2,4.5,20.5}}} = 6.158.$$

Appendix F Statistical Tables

Table F1 Critical values of $|q^*|_{J,\nu}$ [$\alpha = .05$ (.10)]

Table F2 Sample size (n) required to control half-width (w) of standardized Tukey confidence intervals on comparisons [$\alpha = .05$ (.10)]

Table F3 Sample size (n) required to control half-width (w) of standardized Bonferroni-t confidence intervals on comparisons [$\alpha = .05$ (.10)]

Table F4 Sample size (n) required to control half-width (w) of standardized Scheffé confidence intervals on comparisons [$\alpha = .05$ (.10)]

Table F5 Gamma (γ) as a function of J and $(1 - \beta)$ for Tukey tests [$\alpha = .05$ (.10)]

Table F6 Gamma (γ) as a function of k and $(1 - \beta)$ for Bonferroni-t tests [$\alpha = .05$ (.10)]

Table F7 Gamma (γ) as a function of ν_1 and $(1 - \beta)$ for Scheffé tests [$\alpha = .05$ (.10)]

Table F1(a) Critical values of $|q^*|_{J,\nu}$ for $\alpha = .05$

ν \ J	2	3	4	5	6	7	8	9
9	2.262	2.792	3.122	3.363	3.552	3.708	3.841	3.956
10	2.228	2.741	3.059	3.291	3.473	3.623	3.751	3.861
11	2.201	2.701	3.010	3.234	3.410	3.555	3.679	3.785
12	2.179	2.668	2.969	3.187	3.359	3.500	3.619	3.723
13	2.160	2.641	2.935	3.149	3.316	3.454	3.570	3.671
14	2.145	2.617	2.907	3.116	3.280	3.415	3.529	3.628
15	2.131	2.598	2.882	3.088	3.249	3.381	3.493	3.590
16	2.120	2.580	2.861	3.064	3.222	3.352	3.462	3.557
17	2.110	2.565	2.843	3.043	3.199	3.327	3.435	3.529
18	2.101	2.552	2.826	3.024	3.178	3.304	3.411	3.504
19	2.093	2.541	2.812	3.007	3.160	3.285	3.390	3.482
20	2.086	2.530	2.799	2.992	3.143	3.267	3.371	3.462
22	2.074	2.512	2.777	2.967	3.115	3.236	3.339	3.427
24	2.064	2.497	2.759	2.946	3.092	3.211	3.312	3.399
26	2.056	2.485	2.743	2.928	3.072	3.190	3.289	3.375
28	2.048	2.474	2.730	2.914	3.056	3.172	3.270	3.355
30	2.042	2.465	2.719	2.901	3.042	3.157	3.254	3.338
35	2.030	2.447	2.697	2.875	3.013	3.126	3.221	3.303
40	2.021	2.434	2.680	2.856	2.992	3.103	3.197	3.277
45	2.014	2.424	2.668	2.841	2.976	3.086	3.178	3.257
50	2.009	2.415	2.658	2.830	2.963	3.071	3.163	3.241
60	2.000	2.403	2.643	2.812	2.944	3.051	3.140	3.218
70	1.994	2.395	2.632	2.800	2.930	3.036	3.124	3.201
80	1.990	2.388	2.624	2.791	2.920	3.025	3.113	3.188
100	1.984	2.379	2.613	2.778	2.906	3.009	3.096	3.171
120	1.980	2.373	2.605	2.770	2.896	2.999	3.085	3.159
∞	1.960	2.344	2.569	2.728	2.850	2.948	3.031	3.102

Table F1(b) Critical values of $|q^*|_{J,\nu}$ for $\alpha = .10$

ν \ J	2	3	4	5	6	7	8	9
9	1.833	2.345	2.660	2.888	3.067	3.214	3.338	3.446
10	1.812	2.313	2.619	2.841	3.015	3.157	3.278	3.382
11	1.796	2.287	2.587	2.804	2.973	3.112	3.230	3.331
12	1.782	2.266	2.560	2.773	2.939	3.075	3.190	3.289
13	1.771	2.248	2.538	2.747	2.910	3.044	3.157	3.254
14	1.761	2.233	2.519	2.725	2.886	3.017	3.128	3.224
15	1.753	2.220	2.503	2.706	2.865	2.995	3.104	3.199
16	1.746	2.209	2.489	2.690	2.847	2.975	3.083	3.176
17	1.740	2.199	2.477	2.676	2.831	2.958	3.064	3.157
18	1.734	2.191	2.466	2.663	2.817	2.942	3.048	3.139
19	1.729	2.183	2.456	2.652	2.804	2.929	3.033	3.124
20	1.725	2.176	2.448	2.642	2.793	2.916	3.020	3.110
22	1.717	2.164	2.433	2.625	2.774	2.895	2.998	3.086
24	1.711	2.155	2.421	2.611	2.758	2.878	2.979	3.066
26	1.706	2.147	2.410	2.599	2.744	2.863	2.963	3.050
28	1.701	2.140	2.402	2.588	2.733	2.851	2.950	3.035
30	1.697	2.134	2.394	2.579	2.723	2.840	2.938	3.023
35	1.690	2.122	2.379	2.562	2.703	2.819	2.915	2.999
40	1.684	2.113	2.368	2.549	2.689	2.803	2.898	2.980
45	1.679	2.106	2.359	2.539	2.677	2.790	2.885	2.966
50	1.676	2.100	2.352	2.531	2.668	2.780	2.874	2.955
60	1.671	2.092	2.342	2.519	2.655	2.765	2.858	2.938
70	1.667	2.087	2.335	2.510	2.645	2.755	2.847	2.926
80	1.664	2.082	2.329	2.504	2.638	2.747	2.839	2.917
100	1.660	2.076	2.321	2.495	2.628	2.736	2.827	2.905
120	1.658	2.072	2.316	2.489	2.622	2.729	2.819	2.896
∞	1.645	2.052	2.291	2.460	2.589	2.693	2.780	2.855

Table F2 Sample size (n) required to control half-width (w) of standardized Tukey confidence intervals on comparisons [$\alpha = .05$ (.10)]

w \ J	2	3	4	5	6	7	8	9
.10	770 (543)	1100 (844)	1321 (1052)	1490 (1211)	1626 (1342)	1740 (1452)	1839 (1547)	1926 (1631)
.15	343 (242)	490 (411)	588 (557)	663 (693)	723 (823)	774 (948)	818 (1070)	857 (1189)
.20	194 (137)	276 (212)	331 (264)	374 (304)	408 (337)	436 (364)	461 (388)	483 (409)
.25	124 (88)	177 (136)	213 (170)	240 (195)	261 (216)	280 (234)	295 (249)	309 (262)
.30	87 (62)	124 (95)	148 (118)	167 (136)	182 (150)	195 (163)	206 (173)	215 (183)
.35	64 (46)	91 (70)	109 (87)	123 (100)	134 (111)	143 (120)	151 (128)	159 (135)
.40	50 (35)	70 (54)	84 (67)	95 (77)	103 (85)	110 (92)	116 (98)	122 (103)
.45	39 (28)	56 (43)	67 (53)	75 (61)	82 (68)	87 (73)	92 (78)	97 (82)
.50	32 (23)	45 (35)	54 (44)	61 (50)	66 (55)	71 (60)	75 (63)	78 (67)
.55	27 (19)	38 (29)	45 (36)	51 (41)	55 (46)	59 (49)	62 (53)	65 (55)
.60	23 (17)	32 (25)	38 (31)	43 (35)	47 (39)	50 (42)	53 (44)	55 (47)
.65	20 (14)	28 (21)	33 (26)	37 (30)	40 (33)	43 (36)	45 (38)	47 (40)
.70	17 (13)	24 (19)	28 (23)	32 (26)	35 (29)	37 (31)	39 (33)	41 (35)
.75	15 (11)	21 (16)	25 (20)	28 (23)	30 (25)	32 (27)	34 (29)	36 (30)
.80	14 (10)	19 (15)	22 (18)	25 (20)	27 (22)	29 (24)	30 (26)	32 (27)

Table F3 Sample size (n) required to control half-width (w) of standardized Bonferroni-t confidence intervals on comparisons [$\alpha = .05$ (.10)]

w \ k	2	3	4	5	6	8	10	15
.10	1006 (770)	1148 (907)	1249 (1006)	1328 (1084)	1394 (1148)	1497 (1249)	1577 (1328)	1725 (1474)
.15	448 (343)	511 (404)	556 (448)	591 (483)	620 (511)	666 (556)	702 (591)	767 (656)
.20	253 (194)	288 (228)	313 (253)	333 (272)	350 (288)	375 (313)	395 (333)	432 (370)
.25	162 (124)	185 (146)	201 (162)	214 (175)	224 (185)	241 (201)	254 (214)	277 (237)
.30	113 (87)	129 (102)	140 (113)	149 (122)	156 (129)	168 (140)	177 (149)	193 (165)
.35	84 (64)	95 (75)	103 (84)	110 (90)	115 (95)	124 (103)	130 (110)	142 (122)
.40	64 (50)	73 (58)	79 (64)	84 (69)	89 (73)	95 (79)	100 (84)	109 (94)
.45	51 (39)	58 (46)	63 (51)	67 (55)	70 (58)	75 (63)	79 (67)	87 (74)
.50	42 (32)	47 (38)	51 (42)	55 (45)	57 (47)	61 (51)	65 (55)	70 (60)
.55	35 (27)	39 (31)	43 (35)	45 (37)	48 (39)	51 (43)	54 (45)	58 (50)
.60	29 (23)	33 (27)	36 (29)	38 (32)	40 (33)	43 (36)	45 (38)	49 (42)
.65	25 (20)	29 (23)	31 (25)	33 (27)	34 (29)	37 (31)	39 (33)	42 (36)
.70	22 (17)	25 (20)	27 (22)	29 (24)	30 (25)	32 (27)	34 (29)	37 (32)
.75	19 (15)	22 (18)	24 (19)	25 (21)	26 (22)	28 (24)	30 (25)	32 (28)
.80	17 (14)	19 (16)	21 (17)	22 (18)	23 (19)	25 (21)	26 (22)	28 (25)

Table F4 Sample size (n) required to control half-width (w) of standardized Scheffé confidence intervals on comparisons [$\alpha = .05$ (.10)]

w \ ν_1	1	2	3	4	5	6	7	8
.10	770 (543)	1200 (923)	1564 (1252)	1899 (1557)	2216 (1849)	2520 (2130)	2815 (2405)	3103 (2674)
.15	343 (242)	534 (411)	696 (557)	845 (693)	986 (823)	1121 (948)	1252 (1070)	1380 (1189)
.20	194 (137)	301 (232)	392 (314)	476 (390)	555 (463)	631 (534)	705 (602)	777 (670)
.25	124 (88)	193 (149)	252 (202)	305 (250)	356 (297)	404 (342)	452 (386)	498 (429)
.30	87 (62)	135 (104)	175 (140)	212 (174)	248 (207)	281 (238)	314 (269)	346 (298)
.35	64 (46)	99 (77)	129 (104)	156 (129)	182 (152)	207 (175)	231 (198)	255 (220)
.40	50 (35)	76 (59)	99 (80)	120 (99)	140 (117)	159 (135)	177 (152)	195 (169)
.45	39 (28)	61 (47)	79 (63)	95 (78)	111 (93)	126 (107)	140 (120)	155 (133)
.50	32 (23)	49 (38)	64 (52)	77 (64)	90 (75)	102 (87)	114 (98)	126 (108)
.55	27 (19)	41 (32)	53 (43)	64 (53)	75 (63)	85 (72)	95 (81)	104 (90)
.60	23 (17)	35 (27)	45 (36)	54 (45)	63 (53)	71 (61)	80 (68)	88 (76)
.65	20 (14)	30 (23)	38 (31)	46 (38)	54 (45)	61 (52)	68 (58)	75 (65)
.70	17 (13)	26 (20)	33 (27)	40 (33)	47 (39)	53 (45)	59 (51)	65 (56)
.75	15 (11)	23 (18)	29 (24)	35 (29)	41 (34)	46 (39)	52 (44)	57 (49)
.80	14 (10)	20 (16)	26 (21)	31 (26)	36 (30)	41 (35)	45 (39)	50 (43)

Table F5 Gamma (γ) as a function of J and $(1 - \beta)$ for Tukey tests [$\alpha = .05$ (.10)]

$1-\beta$ \ J	2	3	4	5	6	7	8
.65	2.345 (2.030)	2.729 (2.438)	2.954 (2.677)	3.113 (2.845)	3.235 (2.974)	3.334 (3.078)	3.416 (3.165)
.70	2.484 (2.169)	2.868 (2.577)	3.093 (2.816)	3.252 (2.984)	3.374 (3.113)	3.473 (3.217)	3.555 (3.304)
.75	2.634 (2.319)	3.018 (2.727)	3.244 (2.966)	3.402 (3.134)	3.524 (3.263)	3.623 (3.367)	3.705 (3.454)
.80	2.802 (2.486)	3.185 (2.894)	3.411 (3.133)	3.569 (3.301)	3.691 (3.430)	3.790 (3.534)	3.873 (3.622)
.85	2.996 (2.681)	3.380 (3.089)	3.605 (3.328)	3.764 (3.496)	3.886 (3.625)	3.985 (3.729)	4.067 (3.816)
.90	3.242 (2.926)	3.625 (3.334)	3.851 (3.573)	4.009 (3.741)	4.131 (3.870)	4.230 (3.974)	4.312 (4.061)
.95	3.605 (3.290)	3.989 (3.697)	4.214 (3.936)	4.373 (4.104)	4.495 (4.233)	4.593 (4.338)	4.676 (4.425)

Table F6 Gamma (γ) as a function of k and $(1 - \beta)$ for Bonferroni-t tests [$\alpha = .05$ (.10)]

$1-\beta$ \ k	2	3	4	6	8	10	15
.65	2.627 (2.345)	2.779 (2.513)	2.883 (2.627)	3.024 (2.779)	3.120 (2.883)	3.192 (2.961)	3.321 (3.098)
.70	2.766 (2.484)	2.918 (2.652)	3.022 (2.766)	3.163 (2.918)	3.259 (3.022)	3.331 (3.100)	3.460 (3.237)
.75	2.916 (2.634)	3.068 (2.803)	3.172 (2.916)	3.313 (3.068)	3.409 (3.172)	3.482 (3.250)	3.610 (3.388)
.80	3.083 (2.802)	3.236 (2.970)	3.339 (3.083)	3.480 (3.236)	3.576 (3.339)	3.649 (3.417)	3.777 (3.555)
.85	3.278 (2.996)	3.430 (3.164)	3.534 (3.278)	3.675 (3.430)	3.771 (3.534)	3.843 (3.612)	3.972 (3.749)
.90	3.523 (3.242)	3.676 (3.410)	3.779 (3.523)	3.920 (3.676)	4.016 (3.779)	4.089 (3.857)	4.217 (3.995)
.95	3.886 (3.605)	4.039 (3.773)	4.143 (3.886)	4.283 (4.039)	4.379 (4.143)	4.452 (4.221)	4.580 (4.358)

Table F7 Gamma (γ) as a function of ν_1 and $(1-\beta)$ for Scheffé tests [$\alpha = .05$ (.10)]

$1-\beta$ \ ν_1	2	3	4	5	6	7	8
.65	2.833 (2.531)	3.181 (2.886)	3.466 (3.174)	3.713 (3.424)	3.934 (3.648)	4.136 (3.852)	4.323 (4.041)
.70	2.972 (2.670)	3.320 (3.025)	3.605 (3.314)	3.852 (3.564)	4.073 (3.787)	4.275 (3.991)	4.462 (4.180)
.75	3.122 (2.820)	3.470 (3.175)	3.755 (3.464)	4.002 (3.714)	4.223 (3.937)	4.425 (4.141)	4.612 (4.330)
.80	3.289 (2.988)	3.637 (3.342)	3.922 (3.631)	4.169 (3.881)	4.390 (4.104)	4.592 (4.308)	4.780 (4.497)
.85	3.484 (3.182)	3.832 (3.537)	4.117 (3.826)	4.364 (4.076)	4.585 (4.299)	4.787 (4.503)	4.974 (4.692)
.90	3.729 (3.428)	4.077 (3.782)	4.362 (4.071)	4.609 (4.321)	4.830 (4.544)	5.032 (4.748)	5.219 (4.937)
.95	4.093 (3.791)	4.440 (4.145)	4.725 (4.434)	4.972 (4.684)	5.193 (4.907)	5.395 (5.111)	5.583 (5.300)

References

Abelson, R.P. (1995). *Statistics as principled argument.* Hillsdale, NJ: Lawrence Erlbaum.

Abelson, R.P. & Prentice, D.A. (1997). Contrast tests of interaction hypotheses. *Psychological Methods*, *2*, 315-328.

APA (2001). *Publication manual of the American Psychological Association* (5th ed.). Washington, DC.

Betz, M.A. & Gabriel, K.R. (1978). Type IV errors and analysis of simple effects. *Journal of Educational Statistics*, *3*, 121-143.

Betz, M.A. & Levin, J.R. (1982). Coherent analysis-of-variance hypothesis testing strategies: A general approach. *Journal of Educational Statistics*, *7*, 193-206.

Bird, K.D. (2002). Confidence intervals for effect sizes in analysis of variance. *Educational and Psychological Measurement*, *62*, 197-226.

Bird, K.D. & Hadzi-Pavlovic, D. (1983). Simultaneous test procedures and the choice of a test statistic in MANOVA. *Psychological Bulletin, 93,* 167-178.

Bird, K.D. & Hadzi-Pavlovic, D. (2003). A new approach to the analysis of factorial effects in two-factor fixed-effects designs. Unpublished manuscript, University of New South Wales, Sydney, Australia.

Bird, K.D., Hadzi-Pavlovic, D. & Isaac, A.P. (2000). *PSY* [Computer software]. Sydney: School of Psychology, University of New South Wales.

Boik, R.J. (1986). Testing the rank of a matrix with applications to the analysis of interaction in ANOVA. *Journal of the American Statistical Association*, *81*, 243-248.

Boik, R.J. (1993). The analysis of two-factor interactions in fixed effects linear models. *Journal of Educational Statistics*, *18*, 1-40.

Cohen, J. (1965). Some statistical issues in psychological research. In B. B. Wolman (Ed.), *Handbook of clinical psychology* (pp. 95-121). New York: McGraw-Hill.

Cohen, J. (1969). *Statistical power analysis for the behavioral sciences.* New York: Academic Press.

Cohen, J. (1988). *Statistical power analysis for the behavioral sciences* (2nd ed.). New York: Academic Press.

Cohen, J. (1994). The earth is round (p<.05). *American Psychologist*, *49*, 997-1003.

Cohen, J., Cohen, P., West, S.G. & Aiken, L.S. (2003). *Applied multiple regression/correlation analysis for the behavioral sciences* (3rd ed.). Mahwah, NJ: Lawrence Erlbaum.

Cook, T.D. & Campbell, D.T. (1979). *Quasi-experimentation: design and analysis issues for field settings.* Chicago: Rand McNally.

Cumming, G., & Finch, S. (2001). A primer on the understanding, use and calculation of confidence intervals that are based on central and noncentral distributions. *Educational and Psychological Measurement*, *61*, 532-575.

Dunlap, W.P., Cortina, J.M., Vaslow, J.B. & Burke, M.J. (1996). Meta-analysis of experiments with matched groups or repeated measures designs. *Psychological Methods*, *1*, 170-177.

Dunnett, C.W. (1955). A multiple comparisons procedure for comparing several treatments with a control. *Journal of the American Statistical Association*, *75*, 789-795.

Glass, G.V. (1976). Primary, secondary, and meta-analysis of research. *Educational Researcher*, 5, 3-8.

Glass, G.V., Peckham, P.D. & Sanders, J.R. (1972). Consequences of failure to meet assumptions underlying the analysis of variance and covariance. *Review of Educational Research*, *42*, 237-288.

Harlow, L.L., Mulaik, S.A. & Steiger, J.H. (Eds.) (1997). *What if there were no significance tests?* Mahwah, NJ: Lawrence Erlbaum.

Harris, R.J. (1994). *ANOVA: An analysis of variance primer*. Itasca, IL: F.E. Peacock.

Harris, R.J. (2001). *A primer of multivariate statistics* (3rd ed.). Mahwah, NJ: Lawrence Erlbaum.

Hays, W.L. (1981). *Statistics* (3rd ed.). New York: Holt, Rinehart & Winston.

Hellier, E., Weedon, B., Adams, A., Edworthy, J. & Walters, K. (1999). Hazard perceptions of spoken signal words. In M.A. Hanson, E.J. Lovesey, & S.A. Robertson (Eds.), *Contemporary Ergonomics* (pp. 158-162). London: Taylor & Francis.

Hoenig, J.M. & Heisey, D.M. (2001). The abuse of power: The pervasive fallacy of power calculations for data analysis. *The American Statistician*, *55*, 19-24.

Hotelling, H. (1931). The generalization of Student's ratio. *Annals of Mathematical Statistics*, *2*, 360-378.

Hsu, J.C. (1996). *Multiple comparisons:Theory and methods*. London: Chapman & Hall.

International Committee of Medical Journal Editors (1997). Uniform requirements for manuscripts submitted to biomedical journals. *Journal of the American Medical Association*, *277*, 927-934.

Kirk, R.E. (1995). *Experimental design: Procedures for the behavioral sciences* (3rd ed.). Pacific Grove, CA: Brooks/Cole.

Klockars, A.J. & Hancock, G.R. (1994). Per-experiment error rates: The hidden costs of several multiple comparison procedures. *Educational and Psychological Measurement*, *54*, 292-298.

Levin, J.R. & Marascuilo, L.A. (1972). Type IV errors and interactions. *Psychological Bulletin*, *78*, 368-374.

Lockhart, R.S. (1998). *Introduction to statistics and data analysis in the behavioral sciences*. New York: Freeman.

Marascuilo, L.A. & Levin, J.R. (1976). The simultaneous investigation of interaction and nested hypotheses in two-factor analysis of variance designs. *American Educational Research Journal*, *13*, 61-65.

Maxwell, S.E. & Delaney, H.D. (1990). *Designing experiments and analyzing data: A model comparison perspective*. Pacific Grove, CA: Brooks/Cole.

McDonald, R.P. (1997). Goodness of approximation in the linear model. In L.L. Harlow, S.A. Mulaik & J.H. Steiger (Eds.), *What if there were no significance tests?* (pp. 199-219). Mahwah, NJ: Lawrence Erlbaum.

Morris, S.B. & DeShon, R.P. (2002). Combining effect size estimates in meta-analysis with repeated measures and independent-groups designs. *Psychological Methods, 7*, 105-125.

Nickerson, R.S. (2000). Significance testing: A review of an old and continuing controversy. *Psychological Methods, 5*, 241-301.

Oakes, W.F. (1986). *Statistical inference: A commentary for the social and behavioral sciences*. New York: Wiley.

O'Brien, R.G. & Kaiser, M.K. (1985). MANOVA method for analyzing repeated measures designs: An extensive primer in multivariate analysis of variance. *Psychological Bulletin, 97*, 316-333.

Olson, C.L. (1974). Comparative robustness of six tests in multivariate analysis of variance. *Journal of the American Statistical Association, 69*, 894-908.

Olson, C.L. (1976). On choosing a test statistic in multivariate analysis of variance. *Psychological Bulletin, 83*, 579-586.

Pruzek, R.M. (1997). An introduction to Bayesian inference and its applications. In L.L. Harlow, S.A. Mulaik & J.H. Steiger (Eds.), *What if there were no significance tests?* (pp. 287-318). Mahwah, NJ: Lawrence Erlbaum.

Reichardt, C.S. (1979). The statistical analysis of data from nonequivalent group designs. In T.D. Cook. & D.T. Campbell), *Quasi-experimentation: design and analysis issues for field settings* (pp. 147-205). Chicago: Rand McNally.

Reichardt, C.S. & Gollob, H.F. (1999). Justifying the use and increasing the power of a *t* test for a randomized experiment with a convenience sample. *Psychological Methods, 4,* 117-128.

Richardson, J.T.E. (1996). Measures of effect size. *Behaviour Research Methods, Instrumentation & Computers, 28,* 12-22.

Rogers, J.L., Howard, K.I. & Vessey, J.T. (1993). Using significance tests to evaluate equivalence between two experimental groups. *Psychological Bulletin, 113,* 553-565.

Rosenthal, R. (1994). Parametric measures of effect size. In H. Cooper & L. Hedges (Eds.), *The handbook of research synthesis*. New York: Russel Sage Foundation.

Rosnow, R.L. & Rosenthal, R. (1989). Definition and interpretation of interaction effects. *Psychological Bulletin, 105*, 143-146.

Roy, S.N. (1953). On a heuristic method of test construction and its use in multivariate analysis. *Annals of Mathematical Statistics, 24*, 220-238.

Scheffé, H. (1953). A method for judging all contrasts in the analysis of variance. *Biometrika, 40*, 87-104.

Schmidt, F.L. (1996). Statistical significance testing and cumulative knowledge in psychology: Implications for training of researchers. *Psychological Methods, 1*, 115-129.

Sedlmeier, P. & Gigerenzer, G. (1989). Do studies of statistical power have an effect on the power of studies? *Psychological Bulletin, 105*, 309-316.

Shadish, W.R., Cook, T.D. & Campbell, D.T. (2002). *Experimental and quasi-experimental designs for generalized causal inference.* Boston: Houghton Mifflin.

Sidak, Z. (1967). Rectangular confidence regions for the means of multivariate normal distributions. *Journal of the American Statistical Association*, *62*, 622-633.

Smithson, M. (2000). *Statistics with confidence*. London: Sage.

Smithson, M. (2003). *Confidence intervals*. Thousand Oaks, CA: Sage.

Steiger, J.H. (1999). *STATISTICA Power Analysis*. Tulsa, OK: StatSoft, Inc.

Steiger, J.H. & Fouladi, R.T. (1997). Noncentrality interval estimation and the evaluation of statistical models. In L.L. Harlow, S.A. Mulaik & J.H. Steiger (Eds.), *What if there were no significance tests?* (pp. 221-257). Mahwah, NJ: Lawrence Erlbaum.

Timm, N.H. (1975). *Multivariate analysis with applications in education and psychology*. Monterey, CA: Brooks/Cole.

Tukey, J.W. (1994). The problem of multiple comparisons. In H.I. Braun (Ed.), *The collected works of John W. Tukey. Vol. VIII: Multiple comparisons: 1948-1983* (pp. 1-300). New York: Chapman & Hall. (Original unpublished paper dated 1953.)

Wilkinson, L. and the Task Force on Statistical Inference, APA Board of Scientific Affairs (1999). Statistical methods in psychology journals: Guidelines and explanations. *American Psychologist*, *54*, 594-604.

Winer, B.J., Brown, D.R. & Michels, K.M. (1991). *Statistical principles in experimental design* (3rd ed.). New York: McGraw-Hill.

Index